SpringerBriefs in Education

We are delighted to announce SpringerBriefs in Education, an innovative product type that combines elements of both journals and books. Briefs present concise summaries of cutting-edge research and practical applications in education. Featuring compact volumes of 50 to 125 pages, the SpringerBriefs in Education allow authors to present their ideas and readers to absorb them with a minimal time investment. Briefs are published as part of Springer's eBook Collection. In addition, Briefs are available for individual print and electronic purchase.

SpringerBriefs in Education cover a broad range of educational fields such as: Science Education, Higher Education, Educational Psychology, Assessment & Evaluation, Language Education, Mathematics Education, Educational Technology, Medical Education and Educational Policy.

SpringerBriefs typically offer an outlet for:

- An introduction to a (sub)field in education summarizing and giving an overview of theories, issues, core concepts and/or key literature in a particular field
- A timely report of state-of-the art analytical techniques and instruments in the field of educational research
- A presentation of core educational concepts
- An overview of a testing and evaluation method
- A snapshot of a hot or emerging topic or policy change
- An in-depth case study
- A literature review
- A report/review study of a survey
- An elaborated thesis

Both solicited and unsolicited manuscripts are considered for publication in the SpringerBriefs in Education series. Potential authors are warmly invited to complete and submit the Briefs Author Proposal form. All projects will be submitted to editorial review by editorial advisors.

SpringerBriefs are characterized by expedited production schedules with the aim for publication 8 to 12 weeks after acceptance and fast, global electronic dissemination through our online platform SpringerLink. The standard concise author contracts guarantee that:

- an individual ISBN is assigned to each manuscript
- each manuscript is copyrighted in the name of the author
- the author retains the right to post the pre-publication version on his/her website or that of his/her institution

Angela Fitzgerald · Kimberley Pressick-Kilborn ·
Reece Mills · Linda Pfeiffer · James Deehan

Contemporary Australian Primary Science Teacher Education

Triumphs, Tensions and Future Directions

 Springer

Angela Fitzgerald ⓘ
Faculty of Education
Higher Colleges of Technology
Abu Dhabi, Abu Dhabi, United Arab
Emirates

Reece Mills ⓘ
Queensland University of Technology
Brisbane, Australia

James Deehan ⓘ
School of Education
Charles Sturt University
Bathurst, NSW, Australia

Kimberley Pressick-Kilborn ⓘ
University of Technology Sydney
and Trinity Grammar School
Sydney, Australia

Linda Pfeiffer ⓘ
School of Education and the Arts
Central Queensland University
Gladstone, QLD, Australia

ISSN 2211-1921   ISSN 2211-193X (electronic)
SpringerBriefs in Education
ISBN 978-981-97-5659-9   ISBN 978-981-97-5660-5 (eBook)
https://doi.org/10.1007/978-981-97-5660-5

This Springer imprint is published by the registered company Springer Nature Singapore Pte Ltd.
The registered company address is: 152 Beach Road, #21-01/04 Gateway East, Singapore 189721,
Singapore

If disposing of this product, please recycle the paper.

Foreword

The quality of primary science education has been the focus of considerable writing and research. A range of concerns have been critically examined including preferred pedagogical approaches, the affordances of authentic learning contexts, the inadequacy of allocated time, low levels of science confidence amongst generalist teachers, increasing complexity of curriculum content, decreasing student engagement, a need to embrace technology, the importance of scientific literacy, and the contribution of STEM education. While not an exhaustive list, these areas of study illustrate a perceived divide between aspirational expectations and the reality of school-based practice.

Understanding primary science education requires careful attention to a holistic picture which is characterised by a range of complexities. Research has not always successfully captured the intricate interconnections of the many areas of influence. The available picture, at best, provides a fragmented impression. As a consequence, a persistent focus has been placed on the inadequacy of generalist primary teachers as science educators. This has sustained a narrative calling for the 'upskilling' of practicing primary teachers and has continually placed the burden for improvement on the shoulders of teachers themselves. Maintaining this focus has also, very effectively, deflected attention from other influences. The relationship between teacher preparation and school-based practice is an important consideration in primary science education, but one which has often not gained deserved attention. While this is a complex space, influenced by changing political agendas and mandated requirements, there is a need to understand how pre-service education aims to build particular perspectives about primary science teaching and learning and the potential of such endeavours to ultimately influence school-based practice.

This book widens the research lens to examine how and why primary science teacher education is enacted in Australian Initial Teacher Education (ITE) programs. The importance of effective ITE is amplified as the authors draw on a range of contemporary research to illustrate the approaches, values and beliefs of primary science teacher educators who strive to provide their students with authentic learning experiences, enhanced science knowledge, and meaningful engagement with science education pedagogies. Tensions are clearly illustrated between the task of creating learning conditions which enable pre-service teachers to become confident and competent science educators and often competing expectations and accountabilities of changing political priorities and agendas. Examining the impact of these changes from the perspective of those whose responsibility it is to prepare undergraduate primary teachers for their professional career provides a new research perspective in primary science education.

Providing effective pre-service teacher education emerges as a challenging endeavour. The book provides insight about how this work is impacted by bureaucratic decisions, often limited by policy and prevailing research agendas and inevitably shaped by requirements to follow mandated curriculum and embed teaching approaches which have gained prominence. Innovation in teacher preparation relies on the ability of teacher educators to demonstrate expertise and resilience as they effectively negotiate complex policy and practice issues. The authors show us how such challenges often make it difficult for pre-service teacher educators to make a positive contribution to their field. Building purposeful partnerships between universities, schools, and community organisations exemplifies this challenge and the many inherent tensions of their work. Teacher educators recognise such collaborations enable their students to experience science in meaningful contexts while developing knowledge of science as a discipline; a particular way of thinking and working. STEM education can also be effectively enacted through such partnerships as the contribution of science knowledge and skills can be easily highlighted, improving the integration of theory and practice through authentic primary education and community-based contexts. A shared professional responsibility for the preparation of teachers can also be nurtured through such collaborations, drawing on the mutual inspiration generated when experienced teachers work alongside pre-service teachers. In these contexts, success can be carefully scaffolded, increasing pre-service teacher confidence and competency. Yet to do so, teacher educators must expertly juggle the pressure of restricted time in terms of limited opportunities for science learning within crowded course requirements and as a consequence limited time to engage with outside partners. This also presents a challenge of nurturing sustained connections with external groups when the service of mutual benefit may be short term in nature. The prevailing course structures and accountabilities magnify the effort required of science educators to achieve the obvious benefits of such experiences. The results are often a testament to the persistence of science educators who aspire to elevate the quality of science education, remediate science disengagement, and empower pre-service teachers to advocate for science education.

The chapters bring into sharp focus how political and educational agendas are inevitably entwined and shape the primary science landscape. The authors critically examine a current course accreditation change in Australian ITE requiring pre-service primary teachers to undertake studies in an area of specialisation. Consideration is given to the associated opportunities and the potential constraints raised by this requirement for pre-service and school-based science education. As the implications of a specialisation in science education remain uncertain for both pre-service teachers and primary science educators, a call for further research highlights a need to understand the nature of expected expertise in this ill-defined space. The book explains how graduates may expect to move into a range of school roles and aligned responsibilities, further challenged by potential confusion between the work of a 'hybrid specialist' and generalist teacher. Difficulties are also anticipated for schools as they work to determine the type of practice architecture required to support such new roles. The impact on primary science education remains unknown yet within this uncertainty, teacher educators must grapple to find meaningful ways to respond to such policy changes.

I find this book provides a new perspective on primary science education. It is a compelling account of research into teacher preparation and the influences which enable or constrain the contribution of this work to the achievement of quality primary science education. The work encourages new ways of thinking and working to produce research that is responsive to educator needs. Most importantly I appreciated how this book disrupts the deficit discourse which prevails. It goes well beyond superficial discussions of teacher inadequacy and the pursuit of technical competence which have been persistent drivers of change in primary science. I valued the focus on contemporary issues and the thoughtfulness displayed as findings were interrogated to gain a deeper insight about the implications for state of primary science education. I trust anyone who has an interest in primary science education will arrive at a similar conclusion through their reading.

Kathy Smith
Associate Professor
RMIT University
Melbourne, Australia

Contents

Chapter 1
What Is the State of Play for Primary Science Teacher Education in Australia?

As a primary science teacher educator, my driving purpose is to support pre-service teachers in becoming confident and competent future teachers of science. While my approach to achieving this has evolved over the last 15 years, my focus on inquiry-based and place-relevant practices has not shifted; instead they have strengthened.
Ange

1.1 Introduction: Why This Book Now?

The spotlight on initial teacher education (ITE) in Australia continues to burn brightly (e.g., Ellis, 2022). This attention remains largely due to competing (and perhaps in some ways, complementary) demands emerging from a significant national ITE reform agenda (Brandenberg et al., 2016) and the foregrounding of a critical teacher shortage in the aftermath of the recent COVID-19 global pandemic (McLean Davies & Watterson, 2022). Through the intersection of these two concerns, a narrative has emerged around ITE policy and practice across the country, landing both on the political radar and in the public consciousness, calling for improvement in the quality and 'classroom readiness' of graduate teachers (Green et al., 2018). Australia does not stand isolated in this particular duality with similar tensions playing out on the worldwide stage (see: LeTendre, 2021; Murray et al., 2019; Welch, 2022). What is largely unknown and unknowable, however, is what happens next to create movement from airing to taking action against these concerns.

While it seems that the global stage might be set for change, what remains unsaid and untested is a more explicit understanding of what such large-scale conversations and considerations might mean (if anything) at a 'grassroots' level. Within the context of ITE, the concept of grassroots can be imagined at a level as granular as a specific course (e.g., a singular unit of study), in terms of disciplinarity (e.g., a learning area such as science or math) or as a more holistic educational perspective (e.g., First nations knowledges). Shifting the focus from the macro to the micro matters because, in reality, change will ultimately be enacted in the development and delivery of ITE curriculum through course or unit-based experiences. In this sense, a call to action is emerging for ITE leaders and academics to engage in meaningful dialogue and subsequent actions within and cross-institutionally to reconceptualise quality ITE programming from the ground up.

This book represents a seizing of that moment by recognising that this point in time provides us with an opportunity to turn our gaze to our respective field within ITE in the Australian context—primary science teacher education—and to surface practical and proactive ways forward in this messy and complex space. In the context of science teacher education and the quality of science learning and teaching in classrooms, there remains a gap in knowledge between what happens in ITE and classroom enactment. While 'classroom-readiness' is certainly a noble intention, the reality is that there is a lack of a holistic understanding of what is actually happening in primary science teacher education nationally, in terms of how these intentions may be achieved. While this subsequent introspection is largely positioned to inform understanding in an Australian setting, it does not do this at the expense of the international body of knowledge or what could be contributed to improving global practice.

The broad intention of this book is to explore how primary science teacher education is enacted in Australian ITE programs and why this is the case. To assist in signposting these intentions across the book as a whole, the following questions are explored and will be foregrounded from a range of perspectives across the six chapters:

- What does primary science teacher education look like in practice in Australia?
- What are the triumphs in the approaches used?
- What are the inherent tensions in these approaches?

Specifically, this chapter seeks to provide an overview of the 'state of play' of primary science teacher education, nationally and internationally, as a means of orienting the reader. This orientation takes place by firstly detailing the nature of primary science teacher education and who works in this field before situating these insights both historically and politically.

1.2   Contextualising Primary Science Teacher Education

The intention of primary science teacher education may be understood, in general terms, as preparing pre-service teachers for science learning and teaching in a primary school context. This broad-brush explanation fails, however, to address the nuance in what it means to do this work and who, in fact, is doing this work, which is readdressed in this section drawing on national and international insights.

1.2.1   The 'Big Picture': International Positioning

As a field of study, research into science teacher education and educators is well established and widely regarded as evidenced through measures such as high citation rates, an extraordinary breadth of quality publications and a number of dedicated, high standard research journals, handbooks and encyclopaedias (e.g., see: Abell et al., 2009; Irez, 2006; Luft & Jones, 2022). This impressive profile does make it challenging, however, to provide a succinct and coherent statement characterising what the field represents. A recent publication by Cofré and colleagues (2015) effectively represented the state of science teacher education on the international stage as being largely dominated by government policies and heavily situated within economic, cultural and political contexts. Ultimately, this finding largely identifies a transcendence of the traditional boundaries of subject matter and science education research. These insights align with our interpretation of the field and connect directly with a longstanding and hard to shift global trend around the mismatch between research and policy in science teacher education (Flores, 2016). Alongside this analysis, we note a significant weighting towards science teacher education research being positioned or focused on secondary as opposed to primary education (Skamp, 2020).

Interestingly, this notable gap was also identified in an edited collection by a prominent Australian researcher, Appleton (2006), well known for his work and contributions to the area of primary science teacher education. The intention of this collection was to draw out the international perspectives underpinning the contemporary issues and practices informing this field of study. While this work was undertaken over 15 years ago and is now somewhat dated, it nonetheless calls on researchers from across the world (though a particularly US-centric focus remains) to document the interpretations and innovations of primary science teacher education at that particular conjecture. Nichols and Koballa (cited in Appleton, 2006) make a particularly significant contribution by calling for more critical conversations about the complex issues that challenge and plague this area. Of interest, Nichols and Koballa (cited in Appleton, 2006) identified four key themes to consider and draw upon to inform such discussions: teacher knowledge, disciplinary depth and breadth, the place of science education in schools, and the key features of primary science education.

While the observations from Nichols and Koballa (cited in Appleton, 2006) are not misplaced, it is important to note that their themes do not push beyond the criticisms and critiques levelled at primary science teacher education for a number of decades, which suggests an opportunity to significantly innovate and interrogate this space. This need for a rethink is equally reflected in the way in which science education, more broadly, is valued and foregrounded internationally over science teacher education. For example, international high-stakes assessment measures, such as the *Trends in International Mathematics and Science Study* (TIMSS) and the *Program for International Student Assessment* (PISA), remain a prominent focus in terms of the work of classroom-based science teachers and national reforms in improving student outcomes and ultimately country rankings in science. The significant role that primary science teacher education plays in lifting student achievement and improving educational outcomes should not be underestimated.

1.2.2   The Local Scene: National Snapshot

While not specifically isolated to the Australian context, experience and research tells us that our primary pre-service teachers tend to enter into the learning area of science as part of their ITE program with high anxiety and low confidence (e.g., Crook & Wilson, 2015; Tytler, 2007). These emotions are typically driven by less than positive experiences of science education in secondary school and an identity characterised by not enjoying and/or not being 'good' at science (Herbert & Hobbs, 2014; Jarrett, 1999). For some pre-service teachers coming to ITE studies later in life, they may not have experienced formal science learning and teaching for a number of decades. Unfortunately, these perceptions of science education are often perpetuated by primary in-service teachers, who may express concerns about their own abilities around teaching science in their classrooms and therefore may not role-model the possibilities inherent in this learning area for pre-service teachers (van Aalderen-Smeets et al., 2011). While this is an unfortunate deficit view and does not represent the experiences, interests and/or expertise of all, primary science teacher education does find itself positioned within this paradigm (Deehan, 2021).

Research has long reported that science education makes up around 3% of the teaching time in an Australian primary school classroom, which equates to less than an hour of instruction per week (Angus et al., 2007; OCS, 2012). As a response to this statistic and in an attempt to boost the amount and quality of science being taught in primary classrooms, some school leaders started advertising for and recruiting teachers in specialist science teaching roles. The role of specialists in Australian primary schools is also playing out in policy in relation to STEM education. Somewhat differently to other countries, STEM education has been garnering a more prominent place in classrooms over the last few years. This trend has extended into large-scale recruitment drives for STEM specialist teachers in primary schools, which has been heavily incentivized by state and national governments alike in Australia (alongside funds for purpose-built STEM education spaces) largely concerned with creating a STEM pipeline for future industry needs (Prinsley & Johnston, 2015). Despite these concerted efforts, there remains a large-scale shortage of specialist science (and STEM) teachers in primary schools with a 'blockage' occurring somewhere between graduating from ITE programs and being recruited (Crook, 2018). The reasons behind why recruiting graduates into specialist science teacher roles is so challenging is not largely understood, but could include a lack of interest, not feeling prepared, or an avoidance of being pigeon-holed into a particular version of teaching.

One solution emerging from a review into ITE almost a decade ago (Craven et al., 2014) was the mandated introduction of specialisations within primary ITE programs. The specialisation was intended to increase teacher capacity and leadership in specific learning areas, such as science, rather than to create specialist teachers per se (see Chap. 4 for a more in-depth exploration). While the impact of this initiative remains to be fully seen and realised, anecdotally primary pre-service teachers are not opting into science or STEM specialisations within their ITE programming.

Large numbers of pre-service teachers are instead choosing to specialise in English (literacy) or Mathematics (numeracy) for preference, which could be perceived as a direct response to Department of Education imperatives and subsequent school agendas. As a result, the teachers that are being recruited into science specialist roles in primary schools are largely those with secondary science backgrounds. This is, perhaps unsurprisingly, similar to the profile of primary science teacher educators, who are largely from secondary education backgrounds (Treagust et al., 2014). The authors of this book are a case in point with only two of us educated and employed as a primary teacher.

1.2.3 A Deep Dive: What Our Project Tells Us About What Is Happening on the Ground

Drawing on survey data collected in late 2019 and early 2020 (see: Fitzgerald et al., 2020b), 29 primary science teacher educators working in ITE across most states and territories in Australia provided insights into their experiences of primary science teacher education. For this chapter, the data has been aggregated and is presented as a narrative to provide a broad-brush illustration of the primary science teacher education community in Australia. An overarching snapshot such as this is useful as it provides a baseline from which to compare and contrast the detail located within specific ITE programs and providers.

Bachelor of Education programs in Australia typically offer two core primary science teacher education units (which in some higher education contexts is referred to as a 'course') with around half offering a specialisation in science and/or science-focused electives. Likewise, Master of Teaching programs usually offer only one core science teacher education unit with science specialization or electives not an option. Regardless of the program, primary science teacher education units are most likely to be offered in a lecture-tutorial (enacted more like a workshop) format for three hours a week over a semester (10–12 weeks).

Of the 29 primary science teacher educators who responded to this survey, their split of professional backgrounds across primary and secondary schooling was 50:50. Around three-quarters of the respondents had a Ph.D. in Education with approximately half holding undergraduate qualifications in science. All of the respondents identified as being active researchers in primary science education. A third of the respondents had spent between 5 and 9 years engaged as a primary science teacher educator with a quarter identifying that they had spent 10–14 years employed as a classroom teacher. Typically, primary science teacher education teams are made up of one to three ongoing academic staff members with multiple sessional staff contributing to teaching in a casual capacity. Of this number, a third of the respondents identified as the only academic team member in primary science teacher education in their School or Faculty. It was not common across the providers for casual marking support to be utilised.

In the Australian context, primary science teacher education is largely informed by constructivism and inquiry-based approaches to science learning and teaching. The intention of primary science education units is predominantly to build the confidence and competence of often reluctant cohorts of pre-service teachers through increasing science knowledge, engaging with science education pedagogies, and providing lived science experiences. It is important to note that the following insights were gathered prior to the global pandemic, which gained international traction from March 2020. On-campus modes of delivery are commonplace across ITE providers, however, around 50% of primary science teacher education units have options for online students. On-campus experiences are developed to replicate primary classrooms through the use of non-specialised teaching spaces and uncomplicated, easy to access resourcing such as basic stationary and household items.

Primary science teacher education units tend to be structured around the key conceptual areas of biology, chemistry, earth and space, and physics, which mirrors the Australian science curriculum documents, both nationally and at state levels. These units balance supporting pre-service teachers to develop in-depth knowledge about the 'big ideas' in science with building their skills and understanding of quality science education pedagogies. Attempts are made to incorporate assessment into science education units that are an authentic representation of and relevant to classroom practice, but in reality, this is weighed up against considerations such as large cohorts, accreditation expectations, time restraints, and a crowded curriculum.

For teacher educators, the joyous aspect of working in primary science teacher education is watching pre-service teachers grow in their enthusiasm, engagement and understanding of science education. They are collectively concerned, however, about the lack of timing and resourcing they have to achieve this growth. In helping to turn this situation around, teacher educators saw significant opportunities in taking more integrative approaches to the teaching of science education to pre-service teachers, including through an increased focus on STEM as a vehicle for change.

1.3 Historically Locating Primary Science (Teacher) Education

In understanding the current state of play in Australian primary science teacher education, we need to backtrack to examine where it has come from and how it has evolved. This section explores some of the global and national trends in primary science teacher education before extracting some of the patterns that reveal what primary science teacher education in Australia looks like now.

1.3.1 Global Trends in Primary Science Teacher Education

To think of science education purely in terms of content knowledge—biology, chemistry, earth and space sciences, and physics—is to take a reductivist view of this learning area. Understandings of science have long moved on from this limited perspective, which is reflected in school science curriculum globally (see: ACARA, 2022a, 2022b; NGSS, 2022). Rather than being strictly viewed in positivist terms (e.g., objective, unchanging, quantifiable), a more contemporary view of science has been largely adopted, which finds space to conceptualise science as subjective, tentative and as existing in qualitative forms (Mazzocchi, 2006). In terms of science education, in both primary and secondary settings, this conceptualisation of science has been most notably captured by the construct known as the nature of science (NOS) (see: Abd-El-Khalick et al., 1998). Hodson (2009) compared the teaching of science to the way an anthropologist teaches about another culture with the similarities involving a group of people holding particular knowledge, language, customs, traditions, practices, attitudes, and values. Essentially, NOS encompasses the characteristics, values and assumptions that scientific knowledge is based on and how it is developed (Science Learning Hub, 2022).

In shifting the ways in which science is understood, another particularly important trend in science education over the last few decades has been the acknowledgement and valuing of alternative ways of knowing science (see: Aikenhead, 1996). This construct embraces First Nations and Indigenous peoples ways of making sense of the world as a form of science knowledge that is holistic in nature and has an interdisciplinary applicability to a range of contexts (e.g., agricultural, medicine, ecology, etc.). In school contexts, teachers are bridging the gap between traditional and western views of scientific knowledge by engaging students in learning and teaching activities that draw on both perspectives as an approach to deeply understand a scientific concept (Sotero et al., 2020). For example, in Canada, this concept is referred to as 'Two-Eyed Seeing' and is an inquiry approach designed to support students (or anyone really) to consider the world and the ways they make sense of it using two lenses: Indigenous and western (Jeffery et al., 2021).

Of particular importance to primary science education has been the notable shift from teacher-centred modes of knowledge delivery to student-driven exploration, discussion and open questioning, which has supported students to feel included and valued as science learners (Bybee, 2006). Inquiry-based approaches to science education have played a significant role in achieving this change in practice (Bybee, 1997). Inquiry-based methods emphasise that student curiosity, observations, problem solving and experimentation lead to critical thinking and reflection about science understandings (European Commission, 2007). To better understand the role of inquiry-based methods, a working group organised by the InterAcademy Panel (InterAcademy Panel, 2006) examined the implementation of *Inquiry-Based Science Education* (IBSE) programmes across different countries. As part of their work, the group identified IBSE as a model for "engaging students in identifying relevant evidence, in critical and logical reasoning about it and in reflection on its interpretation" (p. 4).

1.3.2   Evolution of Primary Science Teacher Education in Australia

About 20 years ago, Appleton and his colleagues (2000) wrote extensively about the development of pre-service primary science teacher education in Australia. Their chapter tracked the changes in this field over the previous 30 years using an historical narrative approach alongside a focus on surfacing some of the innovations to illustrate what 'contemporary' programming in primary science teacher education looks like. Without documenting all of the changes from the preceding decades, some recent notable updates that potentially impact the current shape of primary science teacher education include:

- The shift from a one-year graduate teacher qualification (e.g., Graduate Diploma of Teaching) to a two-year program requirement (e.g., Master of Teaching) (Cervini, 2016);
- The primary years of schooling in all Australian states and territories encompassing Years 1–6 (students aged six to 12 years) with Western Australia, the Northern Territory and Queensland, respectively, the last states to shift Year 7 into the secondary schooling system;
- The inclusion of 'hurdles' that pre-service teachers are required to clear either prior to enrolling or graduating from their ITE program, such as:

  - The assessment of non-academic capabilities via the completion of a written submission or online simulation as a Non-Academic Requirement for Teacher Entry (NARTE),
  - Meeting the literacy and numeracy benchmarks through the nationally administered Literacy And Numeracy Test for Initial Teacher Education (LANTITE), and
  - Completion of a capstone project known as a Teaching Performance Assessment (TPA), which is assessed against the graduate-level Australian Professional Standards for Teachers (APST).

- The introduction of the national Job Ready Graduate reforms in 2020, which included a 42% reduction in the cost to complete an ITE program; and
- The introduction of a national Australian Curriculum in 2014, which has not been taken up by all the states (Victoria and New South Wales most notably as the most populated Australian states), with its 8th version of the national science curriculum delivered in 2022 and the 9th version to be phased in over the next few years.

Alongside these significant structural shifts in the education system in Australia and the politically influenced changes in ITE, the requirement for all ITE programs to be externally accredited by the relevant state-based regulators every five years was introduced in 2011 (AITSL, 2022). For primary ITE, a minimum requirement was set at the delivery of one science teacher education course per program. Of the 37 accredited ITE providers in Australia, there are 134 primary programs that include at least one primary science teacher education course with most investing in more than the minimum (e.g., two courses) (AITSL, 2022).

One other notable change connected with the introduction of a national curriculum in Australia that holds particular relevance and importance for science education was the inclusion of a strand referred to as 'science as a human endeavour' (SHE) (Aldous et al., 2020). The addition of the SHE strand to school science was a significant shift in principle and signalled a more contemporary way of conceiving what science is. This change required science teachers to cover curriculum with their students that enabled them to more deeply understand the complexities and nuances of the nature of science (NOS). Importantly, this required a move beyond the tried-and-true experimental approach often perpetuated in school science that follows a recipe-like set of steps resulting in a very scaffolded laboratory report. Equally, the introduction of SHE required primary and secondary science teacher educators alike to shift their practices (Aldous et al., 2020).

1.3.3 Patterns in Our Data: What Primary Science Teacher Education Looks Like Now

In revisiting the earlier reported survey data (see: Fitzgerald et al., 2020b), four key trends emerged from the 29 participants regarding their perspectives about what primary science teacher education looks like now.

Firstly, most of the educators made reference to the mandated five-yearly cycle of accreditation and the subsequent impact on ITE programming. Rather than considering this process a burden, it was perceived as enabling the regular review of primary science teacher education courses, which led to their update and alteration. The educators viewed this as an opportunity to improve the quality of their courses and to ensure currency both in terms of the integration of up-to-date practices and theory in primary science education.

Secondly, many of the educators noted a shift in the way that primary science teacher educators thought and spoke about science concepts and content in their courses. The phrase, the 'big ideas' of science education (Harlen, 2010), had gained traction and was being used as a way to organise and structure learning within primary science teacher education. This approach to understanding science was borne out of concerns that many students did not find their science education interesting or relevant to their lives. This large-scale project supported by the Association for Science Education in the United Kingdom resulted in the development of 14 'big ideas', which included conceptual statements like: "objects can affect other objects at a distance" and "all material in the Universe is made of very small particles" (Harlen, 2010, p. 8).

Thirdly, a number of the educators identified that theme-based approaches to planning and implementing units of work in primary schools were having a particular resurgence. This interdisciplinary approach is considered by primary school teachers as being a more effective and efficient means of teaching that associates different learning areas together often under the banner of a particular and often broad topic, such as sustainability or cycles (Deneme & Ada, 2012). Interdisciplinarity, in this case, can mean combining elements of STEM, such as science and mathematics, or foregrounding particular elements, like design-based thinking. Importantly, socio-scientific issues (SSI) were also being noted by the educators as having a more prominent role in unifying and making relevant the primary science curriculum through an increase of engagement in real-world problems (Garrido Espeja & Couso Lagaron, 2015). This interdisciplinarity in science learning and teaching was considered as enabling students to become well-supported, creative and better-informed citizens. From the perspective of the survey respondents, these classroom-based practices had a significant influence on their own approach to primary science teacher education.

Fourth and finally, the educators noted a change from largely examining content and pedagogical knowledge in their courses as separate entities to a focus on science pedagogical content knowledge (PCK) in primary science teacher education. The focus on science PCK was largely about developing future primary teachers of science with an understanding of the specific approaches and strategies that best support the learning of science knowledge, skills and practices. For many of the educators, however, they recognised that in primary science teacher education, in particular, there was still a need to improve the science content knowledge of the pre-service teachers before they could truly start to develop their science PCK. In primary education, pedagogical knowledge development was viewed as less of an issue as it was readily embedded across all courses and significantly highlighted when in schools undertaking professional experience.

1.4 Unpacking the Policy Climate Influencing Primary Science Teacher Education

With primary science (teacher) education now located in terms of global and local trends, there is a developing sense of where primary science teacher education in Australia has been and where it is headed. This section further contextualises the landscape by delving more into how the policy context, especially in relation to STEM, shapes the ways in which primary science teacher education is currently being understood and enacted.

1.4.1 Ebbs and Flows in International Policy

It is hard to ignore the global presence of STEM (science, technology, engineering, and mathematics) and its influence on the ways in which we understand and practice science education (Martin Paez et al., 2019). Regardless of how you define this interdisciplinary construct, the growing focus on STEM professions and the future-oriented role of STEM in the workforce is becoming ever sharper and more prominent (Fitzgerald et al., 2020a). To illustrate this, consider the following insights from the USA (NSB, 2021):

- By the end of 2018, there were more than 1.2 million job openings in STEM-related occupations;
- Only 16% of bachelor degrees obtained by 2020 specialised in STEM-focused disciplines; and
- Within the next decade, 80% of jobs will require technology skills and expertise.

These statements become even more sobering for educators when considered in light of this quote from Alexis Ringwald, co-founder and CEO of *LearnUp*, "65% of today's kids will end up doing jobs that haven't even been invented yet" (World Economic Forum, 2021, para 1). The alignment of the above-mentioned knowns with this unknown is providing the impetus for STEM to have a presence in school-based education. This is at odds, however, with what is happening at the chalk face in schools. STEM is not an acknowledged component of the prescribed curriculum in many parts of the world. Regardless, there is a global policy push for space to be found to accommodate and integrate STEM learning and teaching into classroom activities. The reality of this imperative is that school-based engagement with STEM capabilities and competencies is typically becoming the responsibility of generalist classroom teachers, the approach used in primary education. This is leaving teachers' questioning what is required to ensure STEM education is enacted in meaningful and authentic ways to equip students with the skills, knowledge and attributes that will be valued and needed to be productive contributors in a STEM-focused future (Gardner et al., 2019). This line of thinking is further unpacked and expanded in Chap. 5.

With the context in mind and an understanding of the kinds of challenges teachers, particularly those working in the sciences, face in preparing their students for an uncertain future, it is worth turning our thoughts to what this might mean for learning both in primary school settings and as part of primary science teacher education. Projecting into the future for both the science and STEM disciplines, it is recognized that a particular set of skills, knowledge and attributes will be required to experience success and be an effective contributor in the workplace as well as in the community at large. With the rise of automation, this success will no longer necessarily be about manual and routine tasks. Instead the focus is shifting to higher-level skills that go way above and beyond what can be achieved through robotics and production lines. These so-called 21st-century (21C) learning skills are fast becoming the focus driving the purpose of education worldwide, which signals a move away from the learning of information to the learning of what to do with and how to apply this information meaningfully (Bybee, 2010).

1.4.2   Policy Drivers in Australia

Australia, like many countries, is stuck between a rock and a hard place in terms of what educational policy makers and governments are seeking to see achieved through a quality science (or STEM) education agenda. On one hand, it is about feeding the so-called STEM workforce pipeline and improving student outcomes as measured through national and international standardised testing mechanisms. And on the other hand, it is about the creation of a citizenry that is well-informed and makes data-informed decisions. Contemporary thinkers in primary science teacher education in this country (e.g., Tytler, 2007) maintain a focus on a push beyond learning that is an attainment of facts by concentrating on moving thinking to deeper levels and bringing to the fore the complexities inherent in knowledge and knowledge sharing (Sheriff, 2019). This is where the introduction of STEM has the potential to have a positive outcome on the education sphere, particularly if the skill set that integrated STEM engenders is the focus. In many educational contexts and jurisdictions, the 21C skills considered central to learners and learning are reduced and refined from the list above to the 4Cs: *collaboration, communication, creativity*, and *critical thinking* (Lamb et al., 2017). These four skills, which are represented within the Australian Curriculum as 'general capabilities' (ACARA, 2022a, 2022b), are integrated within and across learning areas to not support their development, but better represent how knowledge is explored and expressed beyond the classroom in real world contexts.

In the science education context, inquiry is an important component that speaks directly to 21C skills as well as having a place within the science curriculum in Australia as a 'science inquiry skills' strand (ACARA, 2022a, 2022b). While inquiry is a word that is commonly used and understood in our daily language, it is often difficult to pin down and define in a pedagogical sense (Cochrane-Smith & Lytle, 1999). This confusion often stems from a sense that engaging in inquiry processes in the classroom is about finding out an answer (usually through searching or researching on the internet), but it is much more than that. To think about inquiry in such simplistic terms is to undersell the value in the learning process. It is not just about accessing an established set of facts to find an answer or following a smooth, well-established path to knowledge. The role of inquiry in the classroom is to spark students' curiosity about a particular problem or issue and ignite in them a genuine desire to want to discover a solution (SSEC, 2022). The role of the teacher in this process is more as a facilitator and guide; it is not up to them to work out the answer or the path to get there. Instead, they should be posing the questions and bringing attention to the problems, then showing students the way by providing access to resources and materials that might assist in this learning journey. Equally, learners have a role to play here too. Rather than being passive receivers of information, engagement in inquiry-based approaches to learning requires learners to be active seekers of knowledge and even more active in working out how to apply this knowledge to generate thoughtfully considered responses (Hopenfenbeck et al., 2022).

1.4.3   Policy in Practice: Emerging Implications from Our Data

The above summary of the national and international policy landscape captures what links the STEM-focused future workforce, 21C learning skills and inquiry approaches to science learning together. This connection is a recognition of the need to think and act differently in a world that faces a number of socially, geographically and environmentally complex and multi-faceted problems. To find solutions to these problems, there is a need to engage students in a different approach to learning science throughout their education, so that they are prepared to be the leaders, thinkers and doers in their uncertain futures. Equally, this need is also applicable to our teaching workforce. Therefore, primary science education teachers are in the driving seat to prepare pre-service teachers for this future.

In light of this policy landscape, the 29 survey respondents (see: Fitzgerald et al., 2020b) raised the following three possible concerns and challenges.

1. *Primary science specialisation*

While a primary specialisation is a mandated requirement for primary ITE programs, it is up to each provider to identify which learning areas they will deliver a specialisation in and in what format. The respondents identified that further clarification was required from the national regulator to clarify the purpose and value of the primary specialisation as well as some guidelines that would guide implementation to bring more value to primary teacher education in general and primary science teacher education specifically.

2. *Role of professional experience and work-integrated learning*

ITE providers are required to embed a minimum number of professional experience days into any given program. A concern raised by the respondents was the reality that many pre-service teachers are not experiencing the learning and teaching of science in primary school classrooms. This disjuncture acts to counter the work happening in primary science teacher education courses to increase the confidence and competence of future teachers of primary science and further widens the gap between the perceptions of university-based 'theory' and school-based 'practice.

3. *Integrated STEM education and what it means for primary science education*

Like school teachers, teacher educators in Australia also experience the curriculum as 'crowded' (Hickey, 2021). The survey respondents acknowledge and work with this constraint in their primary science education courses, but raised concern around the addition of integrated STEM into this learning area. Integrated STEM education was viewed by the respondents as having different content and pedagogical considerations to primary science education. The concern is that the competing demands of covering science and STEM within the limitations of a semester potentially acts to diminish solid foundational understanding and skills in both areas.

1.5   Conclusion: What Is the State of Play?

With thanks to a declared national teacher shortage and a continued focus on reform, ITE in Australia is well and truly on the policy and public radar. Within this landscape, primary science teacher education finds itself in an interesting educational time period, which nudges this field closer to a critical juncture of decision-making in terms of which way to go next. This chapter captures the tension that while on one hand the state of play suggests that the future is rich with opportunity and possibilities for innovation, on the other hand there are significant and complex policy and practice issues to negotiate and navigate. For example, global high-stakes standardised testing regimes are narrowing the focus around what matters in science education and pressuring teachers into a version of science education that aligns with 'the test'. Alongside this reality, however, are widespread changes internationally in how school science is conceptualised to value evolving understandings about the nature of science and the value of alternative ways of knowing science. Nationally, the introduction of a regular accreditation cycle for ITE programs provides a mechanism for quality and change in science education in response to contemporary research and industry insights, but the STEM agenda is acting to squeeze an already crowded science education curriculum. At a local level, primary science teacher educators report achieving shifts in pre-service teachers' confidence and competence as future teachers of science with the introduction of the primary specialisation assisting in this cause, but the lack of opportunities for pre-service teachers to experience science learning and teaching in practice are acting to counteract this good work.

This 'see-sawing' between the triumphs and tensions in primary science teacher education can equally be as exhausting as it is exhilarating. What this review tells us is that there is an active and engaged cohort of primary science teacher educators working in ITE in Australia. They are seeking to make a positive contribution to their field but are challenged by a discordant policy landscape, a research agenda that favours secondary science education, and by often being the lone voice for primary science teacher education in their institutions. Importantly, however, they are seeking to engage with the affordances presented by the accreditation process, the embedding of general capabilities in curriculum, and the opening up of interdisciplinary possibilities to better highlight the relevance of authentic science education. So, where to from here? From this scoping exercise, it is hoped that the groundwork has been laid for more critical conversations to emerge that interrogate the spaces inhabited by primary science teacher education and that these robust discussions lead to more innovation in relation to both practice and research. The following chapters intend to continue to prompt and provoke these different ways of thinking. With the initial ground set in terms of establishing a broad brush understanding of the primary science teacher education landscape in Australia, the next chapter seeks to further explore and articulate what the purpose of primary science teacher education is and how that is enacted in this context.

References

Adb-El-Khalick, F., Bell, R. L., & Lederman, N. G. (1998). The nature of science and instructional practice: Making the unnatural natural. *Science Education, 82*(4), 417–436. https://doi.org/10. 1002/(SICI)1098-237X(199807)82:4%3c417::AID-SCE1%3e3.0.CO;2-E

Abell, S., Park Rogers, M. A., Hanuscin, D. L., Lee, M. H., & Gagnon, M. J. (2009). Preparing the next generation of science teacher educators: A model for developing PCK for teaching science teachers. *Journal of Science Teacher Education, 20*(1), 77–93. https://doi.org/10.1007/s10972-008-9115-6

ACARA. (2022a). *Australian curriculum: Science (version 8.4).* https://www.australiancurriculum. edu.au/f-10-curriculum/science/

ACARA. (2022b). *Australian curriculum: General capabilities (version 8.4).* https://www.austra liancurriculum.edu.au/f-10-curriculum/general-capabilities/

Aikenhead, G. S. (1996). Science education: Border crossing into the subculture of science. *Studies in Science Education, 27*, 1–51. https://doi.org/10.1080/03057269608560077

Aldous, C., Cornelius-Bell, A., & Sesterka, A. (2020). *Development of pre-service teachers' conceptualisation of science as a human endeavour: Bridging the gap case study* [Paper presentation]. Australasian Science Education Research Association [ASERA] Conference 2020, Wollongong, Australia.

Angus, M., Olney, H., & Ainley, J. (2007). *In the balance: The future of Australia's primary schools.* Australian Primary Principals Association.

Appleton, K. (Ed.). (2006). *Elementary science teacher education: International perspectives on contemporary issues and practices.* Taylor & Francis.

Appleton, K., Ginns, I. S., & Watters, J. J. (2000). The development of preservice elementary science teacher education in Australia. In S. K. Abell (Ed.), *Science teacher education: An international perspective* (pp. 9–29). Springer. https://doi.org/10.1007/0-306-47222-8_2

AITSL. (2022). *Accreditation of initial teacher education programs in Australia: Standards and procedures.* https://www.aitsl.edu.au/tools-resources/resource/accreditation-of-initial-tea cher-education-programs-in-australia---standards-and-procedures

Brandenberg, R., McDonough, S., Burke, J., & White, S. (2016). Teacher education research and the policy reform agenda. In R. Brandenburg, S. McDonough, J. Burke, & S. White (Eds.), *Teacher education: Innovation, intervention and impact* (pp. 1–19). Springer. https://doi.org/ 10.1007/978-981-10-0785-9_1

Bybee, R. W. (1997). *Achieving scientific literacy: From purposes to practices.* Heinemann.

Bybee, R. W. (2006). Boosting science learning through the design of curriculum materials [paper presentation]. In *Australian council for educational research (ACER) conference.*

Bybee, R. W. (2010). Advancing STEM education: A 2020 vision. *Technology and Engineering Teacher, 70*(1), 30–35.

Cervini, E. (2016, May 10). Goodbye DipEd: The long and expensive road leading to teaching today. *The Sydney Morning Herald.* https://www.smh.com.au/education/goodbye-diped-the-long-and-expensive-road-leading-to-teaching-today-20160509-goppgy.html

Cochran-Smith, M., & Lytle, S. (1999). Relationships of knowledge and practice: Teacher learning in communities. *Review of Research in Education, 24*, 249–305. https://doi.org/10.2307/116 7272

Cofré, H. L., Vergara, C., Gonzalez-Weil, C., Santibáñez, D., Ahumada, G., Furman, M., Podesta, M. E., Camacho, J., Gallego, R., & Pérez, R. (2015). Science teacher education in South America. *Journal of Science Teacher Education, 26*, 45–63. https://doi.org/10.1007/s10972-015-9420-9

Craven, G., Beswick, K., Fleming, J., Fletcher, T., Green, M., & Jensen, B., et al. (2014). *Action now: Classroom ready teachers.* Retrieved February 21, 2018 from https://docs.education.gov. au/system/files/doc/other/action_now_classroom_ready_teachers_print.pdf

Crook, S. (2018). Fixing the shortage of specialist science and maths teachers will be hard, not impossible. *The Conversation.* https://theconversation.com/fixing-the-shortage-of-specialist-sci ence-and-maths-teachers-will-be-hard-not-impossible-99651

Crook, S., & Wilson, R. (2015). Five challenges for science in Australian primary schools. *The Conversation*. https://theconversation.com/five-challenges-for-science-in-austra lian-primary-schools-42413

Deehan, J. (2021). Primary science education in Australian universities: An overview of context and practice. *Research in Science Education*. https://doi.org/10.1007/s11165-021-10026-6

Deneme, S., & Ada, S. (2012). On applying the interdisciplinary approach in primary schools. *Social and Behavioural Sciences, 46*, 885–889. https://doi.org/10.1016/j.sbspro.2012.05.217

Ellis, V. (2022). Initial teacher education: With the profession in crisis, let's not waste the chance for change. *Lens*. https://lens.monash.edu/@education/2022/07/11/1384854/initial-teacher-edu cation-dont-waste-the-crisis

European Commission. (2007). *Science education now: A renewed pedagogy for the future of Europe*. Office for Official Publications of the European Communities.

Fitzgerald, A., Haeusler, C., & Pfeiffer, L. (Eds.). (2020a). *STEM education in primary classrooms: Unravelling contemporary approaches in Australia and New Zealand*. Routledge. https://doi. org/10.4324/9780429277689

Fitzgerald, A., Pressick-Kilborn, K., & Mills, R. (2020b). Primary teacher educators' practices in and perspectives on inquiry-based science education: Insights into the Australian landscape. *Education 3–13: International Journal of Primary, Elementary and Early Years Education, 49*(3), 344–356. https://doi.org/10.1080/03004279.2020.1854962

Flores, M. A. (2016). Teacher education curriculum. In J. Loughran & M. L. Hamilton (Eds.), *International handbook of teacher education*. Springer. https://doi.org/10.1007/978-981-10-036 6-0_5

Gardner, K., Glassmeyer, D., & Worthy, R. (2019). Impacts of STEM professional development on teachers' knowledge, self-efficacy, and practice. *Frontiers in Education, 4*(26). https://doi.org/ 10.3389/feduc.2019.00026

Garrido Espeja, A., & Couso Lagaron, D. (2015). Socio-scientific issues (SSI) in initial training of primary school teachers: Pre-service teachers' conceptualization of SSI and appreciation of the value of teaching SSI. *Social and Behavioral Sciences, 196*, 80–88. https://doi.org/10.1016/j. sbspro.2015.07.015

Green, C., Eady, M., & Andersen, P. (2018). Preparing quality teachers: Bridging the gap between tertiary experiences and classroom realities. *Teaching and Learning Inquiry, 6*(1). https://doi. org/10.20343/teachlearninqu.6.1.10

Harlen, W. (2010). *Principles and big ideas of science education*. https://www.ase.org.uk/bigideas

Herbert, S., & Hobbs, L. (2014). School-based approaches to pre-service primary science teacher education resulting in gains in confidence [paper presentation]. In *Australian teacher education association [ASERA] conference*.

Hickey, C. (2021). *A 'crowded curriculum'? Sure, it may be complex, but so is the world kids must engage with*. The Conversation. https://theconversation.com/a-crowded-curriculum-sure-it-may-be-complex-but-so-is-the-world-kids-must-engage-with-157690

Hopenfenbeck, T. N., Denton-Calabrese, T., Johnston, S. K., Scott-Barrett, J., & McGrane, J. A. (2022). *Facilitating curiosity and creativity in the classroom: An international multi-site video study*. https://www.ibo.org/globalassets/new-structure/research/pdfs/oucea-full-rep ort-creativity-and-curiosity.pdf

Hodson, D. (2009). *Teaching and learning about science: Language, theories, methods, history, traditions and values*. Sense Publishers. https://doi.org/10.1163/9789460910531

InterAcademy Panel. (2006). *Report of working group on international collaboration in the evalua-tion of inquiry-based science education (IBSE) programs*. InterAcademy Panel on International Issues.

Irez, S. (2006). Are we prepared? An assessment of preservice science teacher educators' beliefs about nature of science. *Science Eudcation, 90*(6), 1113–1143. https://doi.org/10.1002/sce. 20156

Jarrett, O. S. (1999). Science interest and confidence among preservice elementary teachers. *Journal of Elementary Science Education, 11*(1), 49–59. https://doi.org/10.1007/BF03173790

Jeffery, T., Kurts, D. L. M., & Jones, C. A. (2021). Two-eyed seeing: Current approaches, and discussion of medical applications. *BC Medical Journal, 63*(8), 321–325.

Lamb, S., Marie, Q., & Doecke, E. (2017). *Key skills for the 21st century: An evidence-based review.* https://inventorium.com.au/wp-content/uploads/2020/06/Key-Skills-for-the-21st-Century-Analytical-Report.pdf

LeTendre, G. K. (2021). *How education reforms can support teachers around the world instead of undermining them.* The Conversation. https://theconversation.com/how-education-reforms-can-support-teachers-around-the-world-instead-of-undermining-them-166528

Luft, J., & Jones, G. M. (Eds.). (2022). *Handbook of research on science teacher education* (1st ed.). Routledge. https://doi.org/10.4324/9781003098478

Martin Paez, T., Aguilera, D., Perales-Palacios, F. J., & Vílchez-González, J. M. (2019). What are we talking about when we talk about STEM education? A review of literature. *Science Education, 103*(4), 799–822. https://doi.org/10.1002/sce.21522

Mazzocchi, F. (2006). Western science and traditional knowledge: Despite their variations, different forms of knowledge can learn from each other. *EMBO Reports, 7*(5), 463–466. https://www.ncbi.nlm.nih.gov/pmc/articles/PMC1479546/, https://doi.org/10.1038/sj.embor.7400693

McLean Davies, L., & Watterson, J. (2022). *Australia's teacher shortage won't be solved until we treat teaching as a profession, not a trade.* The Conversation. https://theconversation.com/australias-teacher-shortage-wont-be-solved-until-we-treat-teaching-as-a-profession-not-a-trade-188441

Murray, J., Swennen, A., & Kosnik, C. (2019). *International research, policy and practice in teacher education: Insider perspectives.* Springer. https://doi.org/10.1007/978-3-030-01612-8

National Science Board. (2021). *The STEM labor force of today: Scientists, engineers, and skills technical workers* [NSB-21-2]. https://ncses.nsf.gov/pubs/nsb20212

Next Generation Science Standards. (2022). *The three dimensions of science learning.* https://www.nextgenscience.org

Office of the Chief Scientist. (2012). *STEM education and the workplace* [Paper series]. https://www.chiefscientist.gov.au/2012/09/stem-education-and-the-workplace

Prinsley, R., & Johnston, E. (2015). *Transforming STEM teaching in Australian primary schools: Everybody's business* [Position paper]. https://www.chiefscientist.gov.au/sites/default/files/Transforming-STEM-teaching_FINAL.pdf

Science Learning Hub. (2022). *Describing the nature of science.* https://www.sciencelearn.org.nz/resources/412-describing-the-nature-of-science

Sheriff, B. (2019). *How exemplary teachers promote scientific reasoning and higher order thinking in primary science* [Doctoral dissertation, Edith Cowan University]. ECU Research online. https://ro.ecu.edu.au/cgi/viewcontent.cgi?article=3248&context=theses

Skamp, K. (2020). Research in science education (RISE): A review (and story) of research in RISE articles (1994–2018). *Research in Science Education, 52*, 205–237. https://doi.org/10.1007/s11165-020-09934-w

Sotero, M. C., Chaves Alves, A. G., Gomes Arandas, J. K., & Trindade Medeiros, M. F. (2020). Local and scientific knowledge in the school context: Characterizarion and content of published works. *Journal of Ethnobiology and Ethnomedicine, 16*(23). https://doi.org/10.1186/s13002-020-00373-5

Smithsonian Science Education Centre. (2022). *What is inquiry-based science?* STEMvisions Blog. https://ssec.si.edu/stemvisions-blog/what-inquiry-based-science

Treagust, D., Won, M., Petersen, J., & Wynne, G. (2014). Science teacher education in Australia: Initiatives and challenges to improve the quality of teaching. *Journal of Science Teacher Education, 26*(10), 81–98. https://doi.org/10.1007/s10972-014-9410-3

Tytler, R. (2007). *Re-imaging science education: Engaging students in science for Australia's future.* Australian Council for Educational Research. https://research.acer.edu.au/cgi/viewcontent.cgi?article=1002&context=aer

Van Aalderen-Smeets, S., van der Molen, J. H. W., & Asma, L. J. F. (2011). Primary teachers' attitudes towards science: A new theoretical framework. *Science Education, 96*(1), 158–182. https://doi.org/10.1002/sce.20467

Welch, A. (2022). *Teacher shortages are a global problem: 'Prioritising' Australian visa won't solve ours.* The Conversation. https://theconversation.com/teacher-shortages-are-a-global-pro blem-prioritising-australian-visas-wont-solve-ours-189468

World Economic Forum. (2021). *The future of jobs and skills.* https://reports.weforum.org/future-of-jobs-2016/chapter-1-the-future-of-jobs-and-skills/?ng_wp_cron=1663348438.521128892 8985595703125#view/fn-1

Chapter 2
What Is the Purpose of Primary Science Teacher Education?

> *Primary science teacher education should be an act of service to schools, teachers, and learners as part of a shared commitment to enhancing scientific literacy. As primary science academics we serve by elevating the quality of science education, remediating science disengagement, and advocating for science education.*
> *James*

2.1 Introduction

The central question underpinning this chapter is *"What is the purpose of primary science teacher education?"*. Despite the simple framing, this is a question without a definitive, fixed answer as science is inextricably linked to the societies within which it is situated. Therefore, this chapter will engage with this question to bring stakeholders towards a shared sense of purpose to inform practice, research, and policy.

The interrelated concepts of science and scientific literacy comprehensively capture many of the possible purposes of science education. In simple terms, a scientifically literate person is able to generalise their scientific knowledge and skills beyond formal educational settings (i.e., Scientific Literacy) while considering broader impacts on their society (i.e., Science Literacy) (Bybee, 1997; Roberts & Bybee, 2014). Such an individual would display traits such as: wide-ranging curiosity, engagement in scientific discourse, reasoned scepticism, scientific process skills, and the capacity to make evidence-based decisions relevant to their life (Goodrum et al., 2001; Mansfield & Reiss, 2020). This purpose can only be achieved through shared commitment and support amongst all primary science education stakeholders. As science education academics in Initial Teacher Education (ITE), we find this driving purpose of scientific literacy to be motivating and reminds us that it's not the end of the world if we come up short. While the purpose of science education in specific contexts can vary, principles of inclusion, societal relevance, and meaningful action are crucial—society is built on science and science is built on society.

A. Fitzgerald et al., *Contemporary Australian Primary Science Teacher Education*, SpringerBriefs in Education, https://doi.org/10.1007/978-981-97-5660-5_2

In this chapter, we will examine the role of primary science teacher education in contributing to the overarching goal of developing science and scientifically literate citizens. We will initially present three guiding principles of primary science teacher education:

1. *Elevate* the quality of science education,
2. *Remediate* science disengagement, and
3. *Advocate* for science education.

These principles will then be contextualised through a discussion of the challenges and opportunities associated with some of the key tensions in primary science teacher education, such as:

- Pedagogical values and accessibility,
- Schools and universities,
- Generalists and specialists,
- Science and STEM, and
- Content and pedagogy.

Although these tensions provide an interesting lens through which to consider primary science teacher education, they should not be taken as simple dichotomies or be viewed as mutually exclusive. Rather, they present some unique challenges and opportunities for pre-service primary academics who must seek to balance these elements. Indeed, it is through tension that opportunities for improvement arise. Finally, conclusions and recommendations to reconcile the principles and tensions will be presented.

2.2   Three Guiding Principles for Primary Science Teacher Education

So, the question remains, how can primary science teacher education contribute to the development of a scientifically literate citizenry? We would argue ITE is instrumental in bridging the divides between the scientific community and the remainder of society by aiding the development of confident and competent primary science teachers. There are three core principles that can maximise the potential of primary science ITE to contribute to the grander goal of a scientifically literate global population, which will now be examined.

2.2.1   Elevate Science Education

Primary science teacher education is instrumental to the ongoing elevation of science education quality by producing confident and competent primary teachers. For this principle, elevate refers to quality rather than societal prominence. To contribute to the elevation of science education quality, primary science academics are tasked with bridging the theory-praxis divide that still challenges primary teachers as they strive to teach in complex settings. (Oates & Seah, 2021). Bringing together seminal socio-cultural educational theories (Nolan & Raban, 2015), burgeoning bodies of science education research (Aubusson et al., 2015, 2019), and expanding frameworks of 'best practice' (Deehan et al., 2022) in ways are practically grounded should be how primary science academics support primary teachers and schools. We find the words of Isaac Asimov to eloquently capture the importance of elevating the quality of primary science education:

Science can be introduced to children well or poorly. If poorly, children can be turned away from science; they can develop a lifelong antipathy; they will be in a far worse condition than if they had never been introduced to science at all. Isaac Asimov (cited in Feller, 2008)

This quote really highlights the potential societal harm for which absent or suboptimal science teaching could be responsible, if only indirectly. Even though Asimov passed away over 30 years ago, his words remain relevant today. In this context, Asimov's words highlight the importance of science education generally and should not be read as a critique of modern science education.

Recent research has shown that primary science academics are particularly focused on providing authentic, student-centred learning experiences in ways that accord with the development of scientific literacy (Deehan, 2021, 2022; Fitzgerald et al., 2021). A recent scoping review of 142 research outputs showed that many of the student-centred approaches that are central to pre-service primary science education, such as student-centred investigations, authentic experiences, and cooperative learning, have overwhelmingly positive impacts on the science content knowledge, skills and attitudes of primary aged students (Deehan et al., 2022). Furthermore, these student-centred approaches are associated with large effect size improvements relative to more traditional, passive learning approaches (Aubusson et al., 2015, 2019; Deehan et al., 2022; Skamp & Preston, 2021). Recent Trends in International Mathematics and Science Study (TIMSS) data from Australia has shown strong, more equitable science achievement amongst Year 4 learners (Thomson et al., 2020a, 2020b). Although the extent to which primary science teacher education has contributed to these positive science education findings is unknown, it is clear that the quality of science education is being elevated. Still there is room for improvement as most OECD nations are scoring below the high international benchmark (550) in Year 4 science TIMSS, which suggests that most primary learners struggle to apply their knowledge to the world beyond the classroom: a central step tenet of scientific literacy (Thomson et al., 2020a).

2.2.2 Remediate Science Disengagement

Primary science teacher education is a vital point of remediation where pre-service teachers can be more fully immersed in science so that they may inspire engagement in their future students. Young people's natural enthusiasm for learning more about their worlds is not always fully fostered through their early years of formal education (ACARA, 2019; Toma et al., 2019). Although large-scale research projects have repeatedly highlighted primary aged students' desire to learn more science (ACARA, 2019; Goodrum & Rennie, 2007; Goodrum et al., 2001), their interest in science learning has shown to dwindle by their later primary school years (Dewitt et al., 2014). Many students, particularly females, grow increasingly disaffected with science as they progress through secondary school (Said et al., 2016), and ultimately many opt out of pursuing science in their non-compulsory years of education (Norton et al., 2018). These issues are now intergenerational (Howitt, 2007) as decades of research has shown that pre-service and in-service teachers are still in need of development in their science engagement and content knowledge (Appleton, 1992, 2003; Pino-Pasternak & Volet, 2020). Science disengagement can be a major threat to the functioning of societies. The following quote is a clear expression of the dangers of inadequate scientific literacy:

> *"Today you put the hairspray on and it's good for 12-minutes. So, if I take hairspray and I spray it in my apartment, which is all sealed, you're telling me that affects the Ozone Layer? I say no way folks, no way!"*. Donald Trump (Schipani, 2016)
>
> As primary science academics, we felt that this quote was particularly stunning. It was surreal to hear a presidential candidate acknowledge publicly that he did not understand the particulate nature of matter and receive uproarious applause. Amidst the many memorable experiences we have collectively lived through in recent times, this seemingly innocuous, throwaway line has always stuck with us.

Many primary science academics seem to relish their role in reinvigorating science learning trajectories (Oates & Seah, 2021). One academic captured the professional focus on science education remediation, *"so we have to culturally bring them back and remedy the negative experiences that they've had"*. There is an ever-expanding body of evidence highlighting the positive impacts that pre-service primary science education has had on cohorts of pre-service primary teachers' science content knowledge (McKinnon et al., 2017), science teaching efficacy beliefs (Deehan, 2017) and overall attitudes towards science (González-Gómez et al., 2019). A comprehensive meta-analysis of 140 research outputs that explored the science teaching efficacy beliefs of undergraduate primary teachers found that many of the student-centred, authentic science teaching practices delivered within primary science ITE programs (Deehan, 2021, 2022; Fitzgerald et al., 2021) were associated with improvements to the science teaching efficacy beliefs of prospective teachers globally (Deehan, 2017). Further to this point, research into 54 German classrooms, with a sample of over 1000 primary aged students, found teacher efficacy to be one of the strongest predictors of students' scientific literacy (Fauth et al., 2019). In short, primary science ITE can remediate science disengagement by developing confident and competent primary science teachers.

2.2.3 Advocate for Science Education

Science education advocacy is an important aspect of primary science teacher education. It is not sufficient for primary teachers to be confident and capable science educators; they must be supported to overcome barriers to the provision of high-quality science education, such as time (Crump, 2005), resource (Rowe & Perry, 2020), socio-economic (Sullivan et al., 2018), and curricular issues (Akar, 2018). Indeed, we should extend the goal of developing confident and competent primary science educators to include a willingness to advocate for science education in the face of systemic challenges. Like scientists, primary science academics and teachers cannot afford to rely on well-meaning citizens to advocate for science education in their stead; as the following quote from Barack Obama demonstrates, even strong external advocacy can inadvertently undercut the broad goal of enhancing scientific literacy:

> *I am confident that if we recommit ourselves to discovery; if we support science education to create the next generation of scientists and engineers right here in America; if we have the vision to believe and invest in things unseen, then we can lead the world into a new future of peace and prosperity.* Obama (2008) This quote captures both the best and worst of public science advocacy. Despite Barack Obama's eloquent articulation of the importance of science to our society, he still places the emphasis on scientists alone, and, in doing so, unintentionally minimizes the importance of scientifically literate and engaged citizens.

The reality is that, despite the aforementioned positive trends in pre-service and in-school primary science education, much of the science taught can place children in passive learning roles that lack authenticity (ACARA, 2019; Banilower et al., 2018; Banilower, 2019). Of even greater concern are reports that science is often taught for less than one hour a week in primary classrooms, well below minimum curricular standards (Tytler et al., 2008). These issues are likely related to complex systemic issues, such as: competing curricular requirements, resourcing limitations and context specific challenges (AITSL, 2021; Rowe & Perry, 2020; Sullivan et al.,

2018). It is imperative that new generations of early career primary teachers are sufficiently prepared to bridge the university-school divides (Anagnostopoulos et al., 2007) to advocate for science education despite the challenges. James Deehan's ongoing research (Deehan et al., 2019, 2020) has found that a strong pre-service primary science teaching program can imbue early career teachers to function as science advocates willing to lead, seek professional development, procure resources, and most importantly, prioritise science in their teaching when faced with negative science socialisation (McKinnon et al., 2017). In short, they can function as scientifically literate role models for their students. The ongoing role of primary science teacher education is to advocate for science education directly and enhance science advocacy amongst primary teachers.

2.3 Tensions in Primary Science Teacher Education: Challenges and Opportunities

The principles of elevation, remediation and advocacy describe the role that primary science teacher education should optimally play in achieving the grander ambition of developing scientifically literate citizens. However, to leave the reader with such principles alone would be an oversimplification. While the key principles should guide the actions of primary science teacher education stakeholders, there are a variety of tensions that can catalyse positive outcomes in primary science ITE. Primary science teacher education is influenced by an array of intersecting factors that influence the principles of elevation, remediation and advocacy in complex ways. After immersing ourselves in academic literature, we reflected as a group of experienced primary science academics and identified a selection of five tensions. These tensions are provided for thought provocation as tensions cannot be universally defined. Below we present our short personalised summaries of the five tensions before they are explored in detail:

- (Pedagogical values and accessibility)—We value student-centred, evidence-based teaching approaches and wish for primary science teacher education to be as widely accessible as possible. However, the shift to more online learning in higher education has created a tension between our pedagogical values and our desire for accessible education.
- (Schools and universities)—Although united in their commitment to students' science learning, tensions have arisen between schools and universities as they have become increasingly divided in both form and function in recent decades. Marketisation, research productivity requirements, and casualised teaching in higher education have made maintaining relationships between schools and universities more challenging. This issue is particularly acute when at the point of pre-service to in-service transition.

- (Generalists and specialists)—There is tension regarding whether primary science should be taught by generalist classroom teachers or specialist science teachers. There are strengths and weaknesses to both approaches that are heavily context dependent.
- (Science and STEM)—There is tension regarding when science disciplinary learning should transition into integrated STEM learning. The challenge is ensuring robust STEM learning experiences do not undermine the disciplinary integrity of science or the other STEM disciplines.
- (Content and pedagogy)—There is a delicate balance between primary science content and pedagogy in the limited time afforded in ITE degree structures. A heavy emphasis on mastering science content (biology, physics, chemistry, geology, and astronomy) in one or two semesters can necessitate more transmissive, atomised learning approaches. Conversely, an emphasis on deeper pedagogies, such as inquiry or project-based learning, can leave some science content areas underdeveloped or missed entirely.

2.3.1 Pedagogical Values and Accessibility

Research has shown that pre-service primary science academics greatly value active, student-centred learning approaches such as inquiry learning, project-based learning and cooperative learning (Deehan, 2021, 2022; Fitzgerald et al., 2021). This accords with the work of Willemse and others (2008) who found that respect, dialogue, commitment, trust, authenticity, vulnerability, joint responsibility, and the identity of the student were key values held by a sample of 54 teacher education academics. It appears that meaningful communication with pre-service primary teachers is a driving influence as the authentic, student-centred approaches that are mainstays in primary science teacher education can be viewed as expressions of professional values that are strongly supported by scholarly evidence. Take for example the commonly utilised practice of problem-based learning; an approach that is perceived to have high moral value in education (Willemse et al., 2008) and has shown to significantly improve learners' science outcomes (Deehan et al., 2022). Ensuring that the core pedagogical values of science education are reflected in ITE programs is not only vital to achieving direct (i.e., competent and confident science educators) and indirect (i.e., a scientifically literate population of globally aware citizens) objectives, it also helps academics to retain their professional commitment despite the pressures of modern university work (Heffernan & Heffernan, 2019).

However, in making primary science teacher education more accessible for a greater variety of people, primary science academics are often required to adopt more asynchronous, teacher-centred pedagogies that can contradict educational theory (Nolan & Raban, 2015), the science education evidence base (Aubusson et al., 2015, 2019; Skamp & Preston, 2021) and their professional values (Deehan, 2022). The existing trend towards online teaching in primary science education was heightened during the Covid-19 pandemic (Deehan, 2021; Fackler & Sexton, 2020). In short, this creates a tension between how pre-service primary science is taught and what it is seeking to teach. This can present a unique contradiction for pre-service primary science academics as they value greater access to education for society but feel online teaching to be contradictory to their personal educational values. One pre-service primary science educator captured a sense of loss as online education has become more prominent in primary science ITE:

> *"We're all kind of at a total loss at how you can teach education online…I don't do online. I refuse to do online actually."* An unnamed ITE science academic (personal communication, 2018)
>
> After a lively and passionate discussion about an academic's many years working as a science educator across different settings, this was her response to my question about online education. The shift in tone was palpable and understandable as she felt her values as an educator were fundamentally undermined by online education. However, she also viewed online education as an important tool for enhancing the accessibility of higher education. For her, and likely many others, this is a tension that can never truly be resolved.

However, with universities increasingly offering blended and online study options to cater for the ever-increasing number of willing distance students (AITSL, 2019; Norton et al., 2018) this tension between pedagogical values and accessibility cannot be ignored. The Covid-19 pandemic brought this tension to the forefront as ITE programs were rapidly forced to innovate and adapt traditionally face-to-face practices (Allen et al., 2020; Ellis et al., 2020). While there are established models and additional affordances associated with online approaches to primary science teacher education (Danaia & Deehan, 2016; Tomas et al., 2015), there are still many pre-service teachers and academics who deeply value the meaningful interactions afforded by face-to-face experiences (Deehan, 2021). As primary science teacher education programs are adapted to cater for more students studying in different ways, it will be essential to ensure the teaching practices remain true to the values of primary science education (i.e., scientific literacy). The worst-case scenario would be that university teaching practices fundamentally contradict the content of ITE programs, leaving academics unfulfilled and pre-service teachers cynical about the role of universities in education. In fact, this issue can exacerbate the "two-worlds problem" wherein ITE programs fail to reflect the realities of school teaching practices (Feiman-Nemser & Buchmann, 1985).

2.3.2 Schools and Universities

Historically, the "two worlds pitfall" has been used to characterise a misalignment between the teaching practices of schools and universities (Anagnostopoulos et al.,

2007). However, as student-centred approaches have become more mainstream, the "two worlds pitfall" should now refer to the systemic differences that further divide schools and universities. The "two-worlds" pitfall is an ever-present cultural challenge in primary science teacher education that we, as primary science academics, often choose to address head on:

> *"Those who can't do, teach. Those who can't teach, teach teachers."* Unknown
> This is a fairly blunt part of a "week one" repertoire designed to start a dialogue between the pre-service teachers and myself about the relationship between universities and schools. I've found it to be a reasonably effective, self-deprecating means of fostering meaningful exchanges at the outset of our primary science subjects.

We argue that the traditional framing of the "two-worlds" pitfall is changing. Historically, the core issue was that the constructivist practices espoused in academic settings did not reflect the prevalence of transmissive instruction in primary school settings (Feiman-Nemser & Buchmann, 1985). Research has shown that this tension has been a central focus of primary science academics during recent decades, with authentic experiences (Deehan, 2021, 2022) and multi-faceted inquiry approaches (Fitzgerald et al., 2021) now all being relatively common as primary science teacher education programs are becoming more oriented toward the goal of scientific literacy. This accords with data from both the U.S. National Survey of Science and Mathematics Education (NSSME) and the Australian National Sample Assessment-Science Literacy (NAP-SL) showing that primary teachers are regularly emphasising deep science conceptual development through active learning, group work, class discussions, and student-centred investigations (ACARA, 2013, 2019; Banilower et al., 2018; Banilower, 2019). Although such trends are positive, they cannot be directly linked to practices in higher education and nor can they override the common sentiment that the accumulated knowledge in academia is not easily accessed by practicing teachers (Confrey et al., 2019). While schools and universities appear to be becoming more aligned pedagogically, the divide may be changing rather than disappearing. That is to say, the new "two worlds" pitfall may no longer relate to differences in the educational values and practices between schools and universities, but rather the distance created between these sectors due to increasingly different requirements and pressures. Indeed, as a consequence of the marketisation in higher education, such as the Dawkins reforms in Australia (Bessant, 2002), the research-centric cultures and casualised working conditions in higher education may be disincentivising the maintenance of school-university relationships (Leathwood & Read, 2022). Despite the challenges, efforts are being made to foster more direct, mutually beneficial school-university partnerships (Hobbs et al., 2018).

2.3.3 Pre-service to In-service Transition

For most early career teachers, the transition from ITE programs to in-service teaching is a high-risk period wherein their professional commitment can either be consolidated or broken (Gallant & Riley, 2017; Weldon, 2018). Research has consistently shown that a significant proportion of recent graduates will leave the profession within their first five years of teaching (Doherty, 2020). Doherty's (2020) systematic review found that attrition amongst early career teachers was associated with workload, financial compensation, work conditions, burnout, and substandard professional leadership. These issues are exacerbated for early career teachers who are less experienced, and by extension, less established in their identities and capabilities as educators. The seemingly inherent challenges associated with the pre-service to in-service transition leave science education particularly vulnerable due to its low status relative to literacy and numeracy (ACARA, 2017). Furthermore, many of the more effective student-centred science teaching practices are resource and time intensive (Deehan et al., 2022) and successful implementation can be impacted by school culture and leadership beyond an early career teacher's immediate control (Settlage et al., 2015). The following anecdote from Dr James Deehan illustrates how fraught university-school transitions can be for early career teachers:

> *"Forget all that S*** you learned at uni, the real learning starts now"*. An unnamed teacher (personal communication, 2010)
> This was uttered to me (James) by a very enthusiastic primary teacher back when I was a pre-service teacher. At the time, it served as an ice-breaker as I couldn't help but laugh at the blunt delivery. I went on to have an excellent, supportive experience teaching at the school, but the words themselves have taken on new meaning for me as a primary science academic—they now serve as a reminder of the risk of becoming detached from the teachers and schools that we must work with and serve in primary science teacher education

It is therefore understandable that teachers at the vulnerable early career stage may choose to emphasise more fundamental aspects, such as classroom management, core numeracy and literacy, and administrative requirements, in ways that minimise their focus on primary science education. While early career primary science challenges require further research, there is some evidence that negative science socialisation can occur at the early career stage in ways that enable these emerging educators to justify the marginalisation of science (Deehan et al., 2020). Indeed, some early career teachers feel less efficacious about their science teaching as they grow to personally understand the systemic challenges to effective primary teaching (Andersen et al., 2004). Contrary to these issues, there is also some evidence that early career teachers are capable of overcoming obstacles to retain the positive science dispositions gained during their ITE to become science leaders (Deehan, 2017; Deehan et al., 2020), with pre-service primary science experiences sometimes being directly attributed as reasons for effective science teaching practice by early career teachers themselves (Deehan et al., 2020). Professional Learning Networks (PLNs) and more informal social media collectives can assist in career transitions (Greenhalgh, 2020). It would be worthwhile for primary science teacher academics to collaborate across universities and jurisdictions to develop undergraduate networks to support teachers as they transition from universities to schools.

2.3.4   Generalists and Specialists

Due to the challenges historically associated with primary science education, there is an ongoing debate as to whether primary science should be taught by generalist primary teachers or specialist science teachers. There are undeniably merits and pitfalls associated with both approaches. Although generalist teachers can struggle with common issues of time, resource limitations, and other structural barriers, alongside possible personal deficiencies in their science content knowledge and confidence (Ardzejewska et al., 2010), they have the capacity to de-silo science education by making authentic connections to other disciplines and school contexts in ways that can foster scientific literacy. Furthermore, an emerging body of evidence suggests that generalist primary science teachers have more positive science dispositions and are becoming increasingly willing to utilise authentic student-centred practices (ACARA, 2013, 2019; Banilower et al., 2018; Banilower, 2019; Deehan et al., 2022). It is also highly unlikely that generalist primary science teachers will be supplanted entirely due the dearth of available science specialists (Fraser et al., 2019). A comprehensive analysis of 30 U.S. primary schools indicated that both generalist and specialist approaches to science education could be equally effective with sufficient resourcing and school-wide commitment (Levy et al., 2016). Indeed, a systematic review of 37 research outputs, covering specialisation in primary science and mathematics education, did not find clear evidence that disciplinary experts had a positive impact on instructional quality or student learning relative to their generalist counterparts (Mills et al., 2020). Subsequent research has also shown no statistically significant differences in year 5 students' science achievement on the basis of specialist and generalist approaches to primary science education (Roach & Wendt, 2022). Indeed, as the following quote demonstrates, even the most dedicated and experienced science specialists can make mistakes or have gaps in their knowledge—meaning there is no perfect solution:

"I just realised that I've been teaching my students the wrong thing for over twenty years". An unnamed science teacher
This was uttered by a seemingly shocked high school science teacher in a professional development session. I was taken by his willingness to admit his shortcoming as something to which we should all aspire. It also serves

as a reminder that everyone can be wrong. More broadly, it speaks to the incredible challenge science teachers face as conduits between scientists and non-scientists across the many discrete science disciplines reflected in science syllabi.

A core challenge for primary science teacher education is that generalist and specialist primary science teaching roles can neither be consistently defined nor conceptualised as fixed career-long roles. Although there is some broad consensus that a specialist science educator should be a disciplinary science expert, there is inconsistency both in what constitutes science disciplinary expertise and its deployment in school settings (Mills et al., 2020). Such fluidity makes it very challenging for primary science teacher education to adapt to either approach. Even government mandates emphasising specialisation are hindered by a limited evidence base and marred by opaque definitions (Bourke et al., 2020). Due to the unclear evidence base and the context specific challenges and opportunities for each school, primary science teacher education should strive to open pre-service teachers to the potential benefits of generalist and specialist science teaching. Educators naturally change roles as they progress through their careers, so it is imperative that all primary teachers and administrators contribute to prioritising science education in their schools. The role of specialist teachers in primary science education is addressed more fully in Chap. 4.

2.3.5   Science and STEM

The relationship between science and interdisciplinary STEM education must be navigated skilfully in ITE as clear complementarities have the potential to undermine the discipline integrity of science, as well as mathematics, technology, and engineering. Yet still, the increasing prevalence of STEM education initiatives (Murphy et al., 2019) and established benefits of STEM education for student learning outcomes (Mildenhall et al., 2019) can make STEM a tantalising prospect. Ostensibly, STEM learning is a logical means of consolidating and extending separate disciplines that are often unified in their commitment to constructivism, student-centred learning, and authenticity (Fowler et al., 2021). However, despite these broad commonalities, Science, Technology, Engineering, and Mathematics each have unique epistemologies, challenges, disciplinary objectives, and definitions of 'best practice' (Davis et al., 2019) that must be overtly addressed for effective STEM integration. While STEM may be envisioned as, and can indeed sometimes be in practice, a powerful consolidation of the skills and knowledge from STEM disciplines in grounded and meaningful ways, it presents considerable pedagogical complexity and resource demands. Unfortunately, STEM can remain a challenging concept for ITE academics:

> *"Amongst the mathematics academics STEM is just science by another name".* An unnamed mathematics education academic (personal communication, 2022)
> *"I've been writing about STEM for over a year, and I still don't really know what it is".* An unnamed science education academic (personal communication, 2020)
> The two statements, delivered years apart during unrelated chats, capture the hostility and confusion towards STEM that can be felt by academics working within the separate disciplines. The prominent position of STEM learning in research and national policy requires explicit action from pre-service primary science academics in terms of how they construct and deliver their science subjects and courses. The challenge of meaningful STEM learning for pre-service teachers can only be made more daunting if STEM views are unsettled and inconsistent amongst primary science academics and those from other STEM disciplines.

The delineation and merging of STEM disciplines must be thoughtfully considered within each educational context unless we are to tacitly accept the dilution of discipline integrity. Such dilution could mean a loss of focus on the foundational goal of developing scientific literacy. STEM has the potential to both amplify the strengths of and exacerbate the challenges associated with primary science teacher education (Deehan, 2021, 2022). For the pre-service teachers themselves, STEM education can either advance the positive science trajectories established in ITE or disrupt their development as science and STEM educators through sheer pedagogical complexity.

In an extensive examination of the state of global STEM education, expert STEM educators from Australia, India, Indonesia, and the USA placed a heavy emphasis on the role of ITE academics in maintaining and developing the STEM pipeline (Sheffield et al., 2018). ITE academics in STEM disciplines need to explicitly reflect on their professional and personal views of STEM education and how these impact their educational choices. Davis and others (2019) note that ITE academics tend towards prioritising the integrity of their specific disciplines. It is imperative that pre-service primary science academics maintain discipline integrity as they work to interpret rather than simply follow STEM directives or mandates. The nebulous nature of STEM may be alleviated by meaningful connections between discipline and STEM subjects to ensure coherent learning experiences for pre-service teachers where the connections amongst disciplines are made as explicit as possible. The educational concept of backward faded scaffolding (Slater et al., 2008) could be a guiding principle in course design as discipline subjects provide more support and explicit teaching to prepare pre-service teachers for more open, independent learning in later STEM subjects. It would also be worthwhile for institutions to support resource intensive authentic, localised STEM experiences as a means of demarcating the transition from disciplinary to STEM learning (Davis et al., 2019). It is hoped that these suggestions would maximise the benefits of STEM learning in ITE by emphasising accordant disciplinary philosophies and aims without diminishing the position of science in already crowded ITE course programs. A more thorough account of STEM in primary science teacher education is provided in Chap. 5.

2.3.6   Content and Pedagogy

The balance between content and pedagogical knowledge in primary science teacher education is a longstanding point of contention. Historically, ITE programs have responded to the sheer number of science domains represented in primary science syllabi by delivering separate science content and pedagogy subjects within their degree pathways (Avery & Meyer, 2012), which can often force a relatively shallow 'week-by-week' shift in targeted content. This type of preference for breadth over depth has been criticised for being anathema to the core principles and evidence base of science education (Bybee, 2014). This is particularly concerning as pre-service teachers are not just shaped by what they learn, they are also influenced heavily by how they learn within their ITE programs (Davis et al., 2019). Despite these issues, accrediting bodies can mandate the separation of content and pedagogy instruction within ITE programs (NESA, 2018). We argue that such separation is not ideal as content and pedagogy are inextricably linked. The following quote and anecdote provided by Dr. James Deehan illustrate our point:

> *"You would need at least five degrees to be fully acquainted with the science content that underpins the primary science curriculum".* Professor David McKinnon (personal communication, 2009)
> This memorable quote was delivered just minutes into the first lecture I ever attended as a first-year pre-service primary teacher. It was designed to establish expectations for the semester to come and orient us all to a focus on deeper science learning. Indeed, my fellow pre-service teachers and I spent nearly all the semester working collaboratively to address our diagnosed gaps in Astronomy content knowledge. For all of us, the hours of research, micro-teaching and reflection were worth it when we achieved mastery outcomes ($\geq$ 70%) in our final exam.

In fact, much of James' research has shown that the thoughtful balance of content and pedagogy within pre-service primary science subjects can positively influence pre-service teachers' science teaching efficacy beliefs and reported practices that are resilient to the challenges of in-service career transition (Danaia & Deehan, 2016; Deehan et al., 2017, 2019, 2020; McKinnon et al., 2017). Any approach to balancing content and pedagogical instruction in primary science teacher education must emphasise depth in ways that maximise contextual strengths; access to discipline experts, relevant local sites and funding opportunities should inform content decisions rather than the impractical goal of covering all science content areas. Educational stakeholders need to be both explicit and realistic about what learning can be accomplished in a three-month university science subject. The often-implicit view that a comprehensive grounding in the earth, physical, chemical, and biological sciences is achievable in around 24 weeks of university study could inadvertently detract from the wider purposes that should drive primary science teacher education.

2.4 Conclusions and Recommendations

We now return to the question introduced at the beginning of the chapter, *"What is the purpose of primary science teacher education?"*. Primary science teacher education must align with the broader goal of developing the science and scientific literacy of citizens; a goal that cannot be universal in its expression as science, and by extension science education, are inextricably bound to serve and advance the societies and communities within which they are situated. There are of course different goals that can drive science education, but science and scientific literacy capture the core elements. In the wider context of science education, primary science teacher education should be driven by three distinct, yet interrelated principles:

1. Elevation of the quality of science education,
2. Remediation of science disengagement, and
3. Advocating for science education.

Merely teaching the content of primary science subjects alone does not fulfil the potential of primary science teacher education. The three principles, alongside other relevant professional, personal, and context information, should inform the short-term and long-term decisions and actions taken by primary science academics. To adhere to these principles, it is imperative that primary science academics and academics actively involve themselves in both the conceptualisation and delivery of mandates and initiatives that impact their practice. The overall quality of science education and the maintenance of strong science education trajectories would be at risk if pre-service primary science academics are not actively guiding science education. While teachers, learners, and scientists should remain the central foci in science education, it would be perilous to overlook the importance of primary science teacher education in connecting and supporting these groups.

In this chapter, we have grappled with (hopefully with at least a modicum of success) several of the key tensions that can influence the capacity of primary science teacher education to fulfil its important purposes. Tensions surrounding pedagogical values and accessibility, schools and universities, generalists and specialists, science and STEM, and content and pedagogy will continue to impact primary science teacher education in complex, but hopefully beneficial ways. These tensions also extend beyond the dichotomies present in this chapter. For example, thoughtfully designed science and STEM subjects could consolidate positive trajectories established in early pre-service primary science subjects by opening pre-service teachers to the possibility of moving between generalist and specialist primary science teaching roles in their careers. Robust science learning is a vital foundation for meaningfully integrated STEM learning. So, with some thought and framing, STEM initiatives could be leveraged to develop more flexible primary science educators more attuned to the scientific literacy of their students.

It is hard to offer concrete recommendations for pre-service primary science education due to the complexity of the field. The standard, fairly cliché, calls for ongoing collaboration, practitioner autonomy and academic consultation on policy all apply here—they are cliché for a reason after all. As a small group of primary science teacher academics, we will offer some recommendations for reflection and critique:

- Primary science teacher academics need to be as explicit as possible about the principles that drive their actions, and these principles should be central in the negotiation between ideal and acceptable circumstances and outcomes. It is particularly important for stakeholders to be realistic about what can be accomplished in ITE—we are aiming to establish solid foundations for teachers to develop in the profession, not creating 'out of the box' or 'set and forget' primary science teachers.
- Primary science teacher education should be extended to support pre-service teachers during their vulnerable transition to in-service status. Although institutional-level partnerships and funded professional development programs may be the gold standard, there is still tremendous value to be found in less formal Professional Learning Networks (PLNs) and social media groups. The maintenance of the personal and professional connections we develop at university can support our graduate teachers and keep us grounded in the school contexts we serve from a distance.
- Backward faded scaffolding, defined as the strategic and gradual removal of learning support as learners become more competent in an educational setting (Slater et al., 2008), should be a core design principle in primary science education subjects and broader ITE courses. Enactment of this principle will require course directors, administrators, and senior primary science academics to foster the necessary cultural and material conditions for more systematic collaboration.
- Authentic, localised STEM experiences could be incorporated into the latter stages of primary science programs to demarcate a transition from disciplinary science learning to possible STEM learning. Not only would this maintain discipline integrity, but it could establish manageable interdisciplinary partnerships that can direct pre-service teachers towards more expansive STEM learning, which may in turn lead to more science advocates and potential science specialists emerging within graduate teacher cohorts.
- Flexibility and depth of science content coverage should be prioritised over breadth in the design and delivery of primary science subjects in ITE. The attributes of academic staff and the opportunities unique to each institution should

drive decisions regarding the inclusion and omission of science content from primary science programs. Efforts to cover all science content are likely to require more shallow, passive approaches to teaching and assessment which undermine the purpose and principles of primary science teacher education.

None of these suggestions are new as many academics are already doing excellent work to enhance the quality of primary science teacher education. It is our hope that if we continue to think and act on the purpose and principles that drive primary science teacher education, we will leave science education in a better place than we found it.

References

Akar, H. (2018). The relationships between quality of work life, school alienation, burnout, affective commitment and organizational citizenship: A study on teachers. *European Journal of Educational Research, 7*(2), 169–180.

Allen, J., Rowan, L., & Singh, P. (2020). Teaching and teacher education in the time of COVID-19. *Asia-Pacific Journal of Teacher Education, 48*(3), 233–236.

Anagnostopoulos, D., Smith, E. R., & Basmadjian, K. G. (2007). Bridging the university–school divide: Horizontal expertise and the "two-worlds pitfall." *Journal of Teacher Education, 58*(2), 138–152.

Andersen, A. M., Dragsted, S., Evans, R. H., & Sørensen, H. (2004). The relationship between changes in teachers' self-efficacy beliefs and the science teaching environment of Danish first-year elementary teachers. *Journal of Science Teacher Education, 15*(1), 25–38.

Appleton, K. (1992). Discipline knowledge and confidence to teach science: Self-perceptions of primary teacher education students. *Research in Science Education, 22*, 11–19.

Appleton, K. (2003). How do beginning primary school teachers cope with science? Toward an understanding of science teaching practice. *Research in Science Education, 33*(1), 1–25.

Ardzejewska, K., McMaugh, A., & Coutts, P. (2010). Delivering the primary curriculum: The use of subject specialist and generalist teachers in NSW. *Issues in Educational Research, 20*(3), 203–219.

Aubusson, P., Schuck, S., Ng, W., Burke, P., & Pressick-Kilborn, K. (2015). *Quality learning and teaching in primary science and technology literature review Sydney.* Association of Independent Schools.

Aubusson, P., Skamp, K., Burke, P. F., Pressick-Kilborn, K., Ng, W., Palmer, T. A., Goodall, A., & Ferguson, J. (2019). *Primary connections: Linking science with literacy stage 6 research evaluation final report.* Prepared for Steering Committee of Primary Connections, Australian Academy of Science. Accessed December 22, 2021 https://opus.lib.uts.edu.au/bitstream/10453/134515/1/Primary%20Connections%20Stage%206%20Evaluation_redacted_report_FINAL.pdf

Australian Curriculum, Assessment and Reporting Authority (ACARA). (2019). *National assessment program—Science literacy year 6 report 2018* (pp. 1–160). ACARA. Retrieved from https://nap.edu.au/docs/default-source/resources/nap-sl-report-2018.pdf?sfvrsn=8737e5e_2

Australian Curriculum, Assessment and Reporting Authority (ACARA). (2013). *National assessment program—Science literacy year 6 report 2012* (pp. 1–118). ACARA. Retrieved from http://www.nap.edu.au/verve/_resources/NAP-SL_2012_Public_Report.pdf

Australian Curriculum, Assessment and Reporting Authority (ACARA). (2017). *Cross-curriculum priorities. Australian Curriculum.* https://www.australiancurriculum.edu.au/f-10-curriculum/cross-curriculum-priorities/

Australian Institute for Teaching and School Leadership (AITSL). (2019). *Initial teacher education: Data report 2019*. https://www.aitsl.edu.au/docs/default-source/research-evidence/ite-data-report/2019/aitsl-ite-data-report-2019.pdf

Australian Institute for Teaching and School Leadership (AITSL). (2021). *Australian teacher workforce data: National teacher workforce characteristics report*. Author. Retrieved on August 16, 2022 from https://www.aitsl.edu.au/docs/default-source/atwd/national-teacher-workforce-char-report.pdf?sfvrsn=9b7fa03c_4

Avery, L. M., & Meyer, D. Z. (2012). Teaching science as science is practiced: Opportunities and limits for enhancing pre-service elementary teachers' self-efficacy for science and science teaching. *School Science and Mathematics, 112*(7), 395–409. https://doi.org/10.1111/ssm.2012.112.issue-7

Banilower, E. R. (2019). Understanding the big picture for science teacher education: The 2018 NSSME+. *Journal of Science Teacher Education, 30*(3), 201–208.

Banilower, E. R., Smith, P. S., Malzahn, K. A., Plumley, C. L., Gordon, E. M., & Hayes, M. L. (2018). *Report of the 2018 NSSME+*. Horizon Research Inc.

Bessant, J. (2002). Dawkins' higher education reforms and how metaphors work in policy making. *Journal of Higher Education Policy and Management, 24*(1), 87–99.

Bourke, T., Mills, R., & Siostrom, E. (2020). Origins of primary specialisation in Australian education policy: What's the problem represented to be? *The Australian Educational Researcher, 47*(5), 725–740.

Bybee, R. W. (1997). *Achieving scientific literacy: From purposes to practices.*

Bybee, R. W. (2014). NGSS and the next generation of science teachers. *Journal of Science Teacher Education, 25*(2), 211–221. https://doi.org/10.1007/s10972-014-9381-4

Confrey, J., McGowan, W., Shah, M., Belcher, M., Hennessey, M., & Maloney, A. P. (2019). Using digital diagnostic classroom assessments based on learning trajectories to drive instruction and deepen teacher knowledge. In D. Siemon, T. Barkatsas, & R. Seah (Eds.), *Researching and using learning progressions (trajectories) in mathematics education* (pp. 75–100). Brill Sense Publishers. https://doi.org/10.1163/9789004396449_004.

Crump, S. (2005). Changing times in the classroom: Teaching as a 'crowded profession.' *International Studies in Sociology of Education, 15*(1), 31–48.

Danaia, L., & Deehan, J. (2016). A model for the creation of cooperative e-learning spaces: Teaching early childhood and primary pre-service teachers how to teach science. *Fusion Journal, 8*, 1–19.

Davis, J. P., Chandra, V., & Bellocchi, A. (2019). Integrated STEM in initial teacher education: Tackling diverse epistemologies. In *Critical, transdisciplinary and embodied approaches in STEM education* (pp. 23–40). Springer, Cham.

Deehan, J. (2022). Primary science education in Australian universities: An overview of context and practice. *Research in Science Education, 52*(6), 1735–1759. https://doi.org/10.1007/s11165-021-10026-6

Deehan, J. (2021). Online education practices and teaching team compositions in Australian pre-service primary science education. *Australian Journal of Teacher Education, 46*(6), 78–97. Retrieved from https://ro.ecu.edu.au/ajte/vol46/iss6/5

Deehan, J., MacDonald, A., & Morris, C. (2022). A scoping review of interventions in primary science education. *Studies in Science Education, 1–43*. https://doi.org/10.1080/03057267.2022.2154997

Deehan, J., Danaia, L., & McKinnon, D. H. (2017). A longitudinal investigation of the science teaching efficacy beliefs and science experiences of a cohort of pre-service elementary teachers. *International Journal of Science Education, 39*(18), 2548–2573. https://doi.org/10.1080/09500693.2017.1393706

Deehan, J., Danaia, L., & McKinnon, D. H. (2020). From students to teachers: Investigating the science teaching efficacy beliefs and experiences of graduate primary teachers. *Research in Science Education, 50*(3), 885–916.

Deehan, J., McKinnon, D. H., & Danaia, L. (2019). A long-term investigation of the science teaching efficacy beliefs of multiple cohorts of pre-service elementary teachers. *Journal of Science Teaching Education, 30*(8), 923–945.

DeWitt, J., Archer, L., & Osborne, J. (2014). Science-related aspirations across the primary–secondary divide: Evidence from two surveys in England. *International Journal of Science Education, 36*(10), 1609–1629.

Doherty, J. (2020). A systematic review of literature on teacher attrition and school-related factors that affect it. *Teacher Education Advancement Network Journal, 12*(1), 75–84.

Ellis, V., Steadman, S., & Mao, Q. (2020). 'Come to a screeching halt': Can change in teacher education during the COVID-19 pandemic be seen as innovation? *European Journal of Teacher Education, 43*(4), 559–572.

Fackler, A. K., & Sexton, C. M. (2020). Science teacher education in the time of COVID-19. *The Electronic Journal for Research in Science & Mathematics Education, 24*(3), 5–13.

Fauth, B., Decristan, J., Decker, A. T., Büttner, G., Hardy, I., Klieme, E., & Kunter, M. (2019). The effects of teacher competence on student outcomes in elementary science education: The mediating role of teaching quality. *Teaching and Teacher Education, 86*, 102882.

Feiman-Nemser, S., & Buchmann, M. (1985). Pitfalls of experience in teacher preparation. *Teachers College Record, 87*(1), 53–65.

Feller, R. J. (2008). An awakening (Part I). *Oceanography, 21*(1), 105–109.

Fitzgerald, A., Pressick-Kilborn, K., & Mills, R. (2021). Primary teacher educators' practices in and perspectives on inquiry-based science education: insights into the Australian landscape. *Education 3–13, 49*(3), 344–356.

Fowler, S., Cutting, C., Kennedy, J., Leonard, S. N., Gabriel, F., & Jaeschke, W. (2021). Technology enhanced learning environments and the potential for enhancing spatial reasoning: a mixed methods study. *Mathematics Education Research Journal, 1–24.*

Fraser, S., Beswick, K., & Crowley, S. (2019). Responding to the demands of the STEM education agenda: The experiences of primary and secondary teachers from rural, regional and remote Australia. *Journal of Research in STEM Education, 5*(1), 40–59.

Gallant, A., & Riley, P. (2017). Early career teacher attrition in Australia: Inconvenient truths about new public management. *Teachers and Teaching, 23*(8), 896–913.

González-Gómez, D., Jeong, J. S., & Cañada-Cañada, F. (2019). Enhancing science self-efficacy and attitudes of pre-service teachers (PST) through a flipped classroom learning environment. *Interactive Learning Environments, 1–12.* https://doi.org/10.1080/10494820.2019.1696843

Goodrum, D., & Rennie, L. (2007). *Australian school science education—National action plan 2008–2012—Volume 1.* Commonwealth of Australia.

Goodrum, D., Hackling, M., & Rennie, L. (2001). *The status and quality of teaching and learning of science in Australian schools.* Department of Education, Training and Youth Affairs.

Greenhalgh, S. P., Rosenberg, J. M., Willet, K. B. S., Koehler, M. J., & Akcaoglu, M. (2020). Identifying multiple learning spaces within a single teacher-focused Twitter hashtag. *Computers and Education, 148*(2020), 1–12.

Heffernan, T. A., & Heffernan, A. (2019). The academic exodus: The role of institutional support in academics leaving universities and the academy. *Professional Development in Education, 45*(1), 102–113. https://doi.org/10.1080/19415257.2018.1474491

Hobbs, L., Campbell, C., & Jones, M. (2018). *School-based partnerships in teacher education: A research informed model for universities, schools and beyond.* Springer Nature.

Howitt, C. (2007). Pre-service elementary teachers' perceptions of factors in an holistic methods course influencing their confidence in teaching science. *Research in Science Education, 37*(1), 41–58.

Leathwood, C., & Read, B. (2022). Short-term, short-changed? A temporal perspective on the implications of academic casualisation for teaching in higher education. *Teaching in Higher Education, 27*(6), 756–771.

Levy, A. J., Jia, Y., Marco-Bujosa, L., Gess-Newsome, J., & Pasquale, M. (2016). Science specialists or classroom teachers: Who should teach elementary science? *Science Educator, 25*(1), 10–21.

Mansfield, J., & Reiss, M. J. (2020). The place of values in the aims of school science education. In D. Corrigan, C. Bunting, A. Fitzgerald, & A. Jones (Eds.), *Values in science education* (pp. 191–209). Springer, Cham. https://doi.org/10.1007/978-3-030-42172-4_12

McKinnon, D. H., Danaia, L., Deehan, J. (June, 2017). The design of pre-service primary teacher education science subjects: The emergence of an interactive educational design model. *Journal of Astronomy & Earth Sciences Education (JAESE), 4*(1), 1–24. https://doi.org/10.19030/jaese. v4i1.9972

Mildenhall, P., Cowie, B., & Sherriff, B. (2019). A STEM extended learning project to raise awareness of social justice in a year 3 primary classroom. *International Journal of Science Education, 41*(4), 471–489.

Mills, R., Bourke, T., & Siostrom, E. (2020). Complexity and contradiction: Disciplinary expert teachers in primary science and mathematics education. *Teaching and Teacher Education, 89*, 103010.

Murphy, S., MacDonald, A., Danaia, L., & Wang, C. (2019). An analysis of Australian STEM education strategies. *Policy Futures in Education, 17*(2), 122–139.

NSW Education Standards Authority (NESA). (2018). *NSW supplementary documentation: Subject content knowledge requirements*. Author. https://educationstandards.nsw.edu.au/wps/wcm/con nect/1bea4323-19a6-4af6-b657-95ae4cea954b/subject-content-knowledge-requirements-pol icy.pdf?MOD=AJPERES&CVID=

Nolan, A., & Raban, B. (2015). Socio-cultural theorists and practical implications. In *Theories into practice: Understanding and rethinking our work with young children and the EYLF* (pp. 29–41). Teaching Solutions.

Norton, A., Cherastidtham, I., Mackey, W. (2018). *Mapping Australian higher education 2018*. Grattan Institute.

Oates, G., & Seah, R. (2021). Learning progressions/trajectories in mathematics and science education: A case for evidence-based curricula reform? Guest Editorial. *Australian Journal of Education, 65*(3), 223–226.

Obama, B. (2008, December 20). Seventh president-elect weekly transition address. *American Rhetoric*. https://www.americanrhetoric.com/speeches/barackobama/barackobamaweeklyt ransition7.htm

Pino-Pasternak, D., & Volet, S. (2020). Starting and staying strong: Pre-service primary teachers' attitudinal profiles towards science learning and their outcomes in an introductory science unit. *The Australian Educational Researcher, 47*, 385–408.

Roach, W., & Wendt, J. L. (2022). An exploration of the use of science specialists and elementary students' science achievement. *Journal of Mathematics and Science: Collaborative Explorations, 18*(1), 90–104.

Roberts, D., & Bybee, R. (2014). Scientific literacy, science literacy and science education. In N. Lederman & S. Abell (Eds.), *The handbook of research on science education* (Vol. 2, pp. 545–558). Routledge.

Rowe, E., & Perry, L. B. (2020). Inequalities in the private funding of public schools: Parent financial contributions and school socioeconomic status. *Journal of Educational Administration and History, 52*(1), 42–59.

Said, Z., Summers, R., Abd-El-Khalick, F., & Wang, S. (2016). Attitudes toward science among grades 3 through 12 Arab students in Qatar: Findings from a cross-sectional national study. *International Journal of Science Education, 38*(4), 621–643. https://doi.org/10.1080/09500693. 2016.1156184

Schipani, V. (2016, May 16). Trump on hairspray and ozone. *Annenberg Public Policy Center*. https://www.factcheck.org/2016/05/trump-on-hairspray-and-ozone/

Settlage, J., Butler, M. B., Wenner, J., Smetana, L. K., & McCoach, B. (2015). Examining elementary school science achievement gaps using an organizational and leadership perspective. *School Science and Mathematics, 115*(8), 381–391.

Sheffield, R. S., Koul, R., Blackley, S., Fitriani, E., Rahmawati, Y., & Resek, D. (2018). Transnational examination of STEM education. *International Journal of Innovation in Science and Mathematics Education, 26*(8), 67–80.

Skamp, K., & Preston, C. (Eds.). (2021). *Teaching primary science constructively* (7th ed.). Cengage.

Slater, S. J., Slater, T. F., & Shaner, A. (2008). Impact of backwards faded scaffolding in an astronomy course for pre-service elementary teachers based on inquiry. *Journal of Geoscience Education, 56*(5), 408–416.

Sullivan, K., Perry, L., & McConney, A. (2018). A comparison of rural educational disadvantage in Australia, Canada, and New Zealand using OECD's PISA. *SAGE Open, 8*(4), 1–12. https://doi.org/10.1177/2158244018805791

Thomson, S., Wernert, N., Rodrigues, S., & O'Grady, E. (2020a). *TIMSS Australia 2019: Highlights.* Australian Council for Educational Research. https://doi.org/10.37517/978-1-74286-616-1

Thomson, S., Wernet, N., Buckley, S., Rodrigues, S., O'Grady, E., & Schmid, M. (2020b). *TIMSS Australia 2019: Volume II school and classroom contexts for learning.* Australian Council for Educational Research. https://doi.org/10.37517/978-1-74286-615-4

Toma, R. B., Greca, I. M., & Orozco Gómez, M. L. (2019). Attitudes towards science and views of nature of science among elementary school students in terms of gender, cultural background and grade level variables. *Research in Science & Technological Education, 37*(4), 492–515.

Tomas, L., Lasen, M., Field, E., & Skamp, K. (2015). Promoting online students' engagement and learning in science and sustainability pre-service teacher education. *Australian Journal of Teacher Education, 40*(11), 79–107. https://doi.org/10.14221/ajte.2015v40n11.5

Tytler, R., Osborne, J., Williams, G., Tytler, K., & Clark, J. C. (2008). *Opening up pathways: Engagement in STEM across the primary-secondary school transition.* Australian Department of Education, Employment and Workplace Relations.

Weldon, P. (2018). Early career teacher attrition in Australia: Evidence, definition, classification and measurement. *Australian Journal of Education, 62*(1), 61–78.

Willemse, M., Lunenberg, M., & Korthagen, F. (2008). The moral aspects of teacher educators' practices. *Journal of Moral Education, 37*(4), 445–466.

Chapter 3
How Do Partnership Practices Characterise Approaches to Australian Primary Science Teacher Education?

As a primary PST myself, as part of my ITE degree I engaged in discipline-specific professional experiences. At a local school, I co-taught a series of maths lessons with peers, and worked responsively with an individual student each week over a semester to support her literacy development. These formative experiences had a huge impact on me, in shaping my professional identity. I looked forward to these regular school visits that were interspersed with my campus-based tutorials. Later as a science teacher educator, I sought to create similar opportunities for my own teacher education students, personally appreciative of the power of university-school partnerships. Kimberley

3.1 Introduction

In this chapter, we provide case snapshots that celebrate and reflect the diversity of practices that Australian primary science teacher educators use to design and deliver their programs to enable 'classroom-ready' graduates (Craven et al., 2014) within a wider reform agenda. The case snapshots capture triumphs through sharing examples of innovative practices that overcome some of the limitations of the traditional block Professional Experience placement model. The key model that is explored in this chapter is the formation of partnerships between universities, schools and other community organisations, such as science museums and field study centres. In each case, the main goal of such partnerships is to promote opportunities for science-specific, embedded professional learning in authentic, contemporary, culturally diverse primary education and community-based contexts.

Some of the emergent tensions from the case snapshots that we explore in this chapter include:

- *'Seeing' science in primary schools*: What opportunities are created in initial teacher education (ITE) programs for pre-service teachers (PSTs) to observe—and

teach—primary science? How do teacher educators innovate in their practice, to overcome obstacles during PSTs' 'regular' Professional Experience placements?
- *Community connections*: How can school and community partnerships be developed and sustained to support contemporary primary science teacher education?

There are three sections to this chapter. First, a brief literature review focuses on partnerships in primary science teacher education, highlighting Australian research studies with some integration of international research. In the second section, three case snapshots from pre-service contexts are presented and discussed, to provide exploration of some of the triumphs and tensions above. In the third and final section, we look to the future, first reflecting on the tensions and challenges to then consider the issue of classroom-readiness. We thus conclude with deliberation on the extent to which ITE actually prepares graduate teachers to teach primary science in contemporary classrooms, with discussion framed in relation to the potential of partnerships.

3.2   Partnerships in Primary Science Teacher Education: A Brief Review of Literature

ITE partnerships that span university and school-based contexts *and* are science-focused have the potential to positively impact teaching practice through providing authentic first-hand experiences for PSTs (Gilbert & Hobbs, 2022; Ma & Green, 2023), to mutually benefit all stakeholders (Jones, 2008; Palmer, 2015; Pressick-Kilborn & Prescott, 2020). In a recent systematic review of the literature, Gilbert and Hobbs (2022) identified a number of reasons why science-focused partnerships make important contributions to teacher preparedness. Opportunities for PSTs to teach science in primary school contexts linked to university science methods courses were highlighted by Gilbert and Hobbs (2022) as distinctive to partnership models, with a view to developing practice reflective of contemporary approaches and enhancing competence and confidence. In building PSTs' self-efficacy for teaching science, partnership initiatives can create early career experiences of success with lasting impact (Kenny et al., 2014; Mansfield & Woods-McConney, 2012; Petersen & Treagust, 2014).

Partnerships also have the potential to disrupt conventional teacher education practice (Ma & Green, 2023). When real classroom experiences form the focus of primary science teacher preparation, rather than contrived examples in lectures or tutorials on campus, PSTs' initial development occurs "within the context of a supportive ecosystem steeped in inquiry-based innovative practice" (Gilbert & Hobbs, 2022, p. 185). Partnership initiatives seek to overcome the so-called theory–practice divide by improving the integration of theory and practice (Neal & Eckersley, 2014). Interestingly, Farrell (2023) conceptualises this space as a 'nexus' rather than a divide, with practice informed by theory and theory informed by practice, thus addressing the disconnect that PSTs can experience between their university learning and traditional block-mode school placements (Deehan, 2022).

High-quality partnerships provide opportunities for PSTs to observe and experience 'good' primary science teaching. While Professional Experience placements are a central component of any ITE program (Palmer, 2015), there are limitations of 'regular' or traditional placements that are impacted by the competence and confidence of the supervising or mentor teacher in relation to science education (Forbes, 2013; Jones & Carter, 2007). Without a competent mentor teacher, primary PSTs are less likely to adapt curriculum materials, and experiment with and enact favoured approaches such as inquiry-based science (Forbes, 2013). The fragmented nature of placement blocks, often dispersed throughout ITE degrees, also pose limitations, as science may not be consistently presented or marginalised in the delivery of the primary curriculum (Deehan, 2022; Jones et al., 2016; Kenny, 2012; Ma & Green, 2023). A challenge thus facing primary science teacher education programs is how to create professional experiences that provide opportunities for PSTs to see, as well as successfully plan for, teach, and evaluate meaningful sequences of science learning that expand their repertoire of pedagogical practices (Deehan, 2022; Fitzgerald et al., 2021). In addressing this challenge, partnerships can create a 'third space' towards improving professional practice (Daza et al., 2021; Farrell, 2023), with the potential to blur boundaries between theory and practice in ways that promote the growth of PSTs' *pedagogical stance* "as they develop a sense of curiosity, wonder, and excitement around science" (Magee & Flessner, 2012, p. 357). Such third spaces bring together academic, practitioner and community knowledge in new and transformative ways that are less hierarchical, with the common goal of teacher learning (Zeichner, 2010) and identity formation (Gilbert & Hobbs, 2022).

School-university partnerships also promote shared professional responsibility for the preparation of teachers, in which in-service primary teachers can play new roles (Ma & Green, 2023; Palmer, 2015). Collaboration amongst in-service teachers and university-based science educators is important in building the pedagogical content knowledge (PCK) of PSTs (Kenny, 2012), with benefits of a supportive cohort approach within a school-based setting (Lowery, 2002) and the potential to ensure greater quality of experience (Kenny, 2012). Furthermore, partnerships also bring potential benefits to in-service teachers' own professional learning, supporting the development of new knowledge and practices (Jones, 2008). In Kenny's (2012) study, co-teaching of science was supported by a triadic structure between in-service and pre-service primary teachers and a university science teacher educator. In-service teachers who initially lacked confidence expressed a positive change in their own motivation and confidence to teach science, as well as gaining insight into their own students and developing new approaches and ideas for teaching (Kenny, 2012). A more recent study by Ma and Green (2023) emphasised that observation of PSTs' teaching inspired in-service teachers to reflect on their own practice, while having PSTs in their classrooms provided additional support and resources for student learning. In other words, there was mutual benefit for all stakeholders in an evolving partnership.

Partnerships and collaborations between universities and informal education environments, such as museums and field study centres, have the potential to engage PSTs in the science education community more widely, preparing PSTs with more holistic views of learning and place-based education (Cooke-Nieves et al., 2022; Griffin, 2012; Lemon & Weller, 2015). Cooke-Nieves et al. (2022) stress the important contribution of professional experiences in informal science learning environments in shaping expansive PST identity formation and interest in science. In such partnership contexts, PSTs can have opportunities to design, implement and evaluate meaningful science lessons specific to the affordances and constraints of the site (Lemon & Weller, 2015). The resources and tools available to teachers in informal science learning environments differ from those in schools, as well as offering unique opportunities for PSTs to develop content knowledge (Cooke-Nieves et al., 2022). In addition, industry partnerships designed to impact future teachers can provide insights into the role of science research and scientists (Gilbert & Hobbs, 2022), as discussed in Chap. 5 of this book.

To conclude, partnerships in primary science teacher education have the potential to create rich opportunities for PST professional learning that overcome some of the limitations of a traditional placement. Through initiatives that provide supportive structures, teacher educators and partner organisations can collaboratively design targeted, scaffolded professional experiences for PSTs in real classrooms or other science learning contexts that are mutually beneficial for all stakeholders, located outside of the traditional placement. In the next section of this chapter, we provide three case snapshots to illustrate and discuss innovative practices in specific primary science ITE partnerships.

3.3 Innovative Practices in Partnerships: Case Snapshots

Strong and effective school-university and university-community partnerships can provide PSTs with innovative discipline-specific professional experiences. In particular, the impact of collaborative planning and teaching (New South Wales Department of Education, 2021) and distinctive, successful experiences of science teaching during ITE are taken into PSTs' future careers. While specific partnership models may vary, in each case initiatives are designed to create authentic opportunities for first-hand learning in real classrooms or settings.

The three case snapshots that follow are each from partnerships between universities and schools or informal learning sites in New South Wales, an Australian state in which the primary Syllabus incorporates *Science and Technology* (New South Wales Education Standards Authority, 2017). For each case snapshot, a brief narrative is presented, followed by analytical commentary with some reference to the wider research literature. While the partnership initiatives described in the case snapshots were primarily designed to enhance PSTs' professional learning in science and technology education, it is important to acknowledge that there were also opportunities created for the professional learning of in-service teachers and university-based teacher educators involved (Ma & Green, 2023; Pressick-Kilborn & Prescott, 2017).

3.3.1 Case Snapshot 1: Embedding PST Science Education in a Primary School Through a Design and Make Day

In the second of two mandatory core Science and Technology education subjects, 75[1] 3rd year Bachelor of Education (Primary) PSTs planned and delivered a whole school 'Design and Make Day' (DM Day) for primary students from 11 classes across Kindergarten (Foundation) to Year 6. The DM Day took over the partner school and ran as an incursion with the activities planned by PSTs replacing regular lessons in the final week of the school term. On the DM Day, small groups of six to 10 primary students were allocated to a teaching team of three to four PSTs with whom they spent the day. The following planning guidelines assisted PSTs to develop a meaningful and smooth-running activity sequence over the day (9:30 am–3 pm, with breaks for morning tea and lunch):

(1) *Before morning tea (90 minutes): Engage the students and gauge prior knowledge, introduce the design brief, guide an embedded investigation, complete the design;*

(2) *Between morning tea and lunch (90 minutes): Dedicate time for making, producing, creating, evaluating; Plan a short presentation of the specific brief, the process and the design solution or product; and*

(3) *After lunch (60 minutes): Share 'show and tell' presentations across groups within each class – celebrate diversity of learning.*

PST co-teaching teams provided the resources needed for their small student group to engage in the activities they had planned. This promoted the use of readily available, inexpensive and recycled materials, or encouraged borrowing of equipment from the university where necessary. On the day, the class teachers and science teacher educators supervised the PSTs, providing guidance and input if requested, however PSTs were encouraged to support one another in their team teaching rather than relying on intervention from the other educators (Fig. 3.1).

In preparation for the DM Day, PSTs formed small teaching teams and were allocated a specific class at the partner school in advance. The school's Science and Technology programs were based on 'Primary Connections'[2] (Australian Academy of Science, nd). To ensure relevance, each PST teaching team aligned the design and

---

[1] This initiative ran for 3 consecutive years with the same partner school. This is an average of the number of participating PSTs across the three years (Pressick-Kilborn & Prescott, 2017).

[2] *Primary Connections* is a comprehensive teaching resource aligned with the *Australian Curriculum: Science* that links science and literacy. Each *Primary Connections* teaching and learning unit is framed using Bybee's (1997) 5Es instructional model.

Fig. 3.1 Photographic image from above of student workshop groups

make sequence they planned with the content focus of their specific class's science learning for that term. The class teachers shared the students' curiosity questions from the early weeks in the term, so that these could inform the PSTs' planning in ensuring meaning and interest. A video conference also was arranged, with children selected from each grade and a teacher at the school to ask them questions about their science learning. PSTs then had the opportunity to ask the students and teacher any questions, which also helped them to refine their planning and build connections with the students' prior knowledge.

In addition, in on-campus workshops leading up to the DM Day, the teacher educators engaged PSTs in a design and create sequence as learners themselves, and provided a variety of frameworks and proformas to support the planning and teaching of design and production tasks. PSTs familiarised themselves with the specific 'Primary Connections' unit that was guiding the science learning for the students in their allocated class, to ensure that their planned activities extended beyond but complemented the content. Finally, in developing their activity sequence for the DM Day, each teaching team had a dedicated small group face-to-face meeting with their Science and Technology tutor to discuss their plans and receive feedback to enable revisions ahead of the DM Day. This discussion also allowed for clarification of their own science understandings in advance and ensured that each PST had nominated to lead the teaching of one of the sessions within the DM Day.

Assessment tasks for the subject were embedded in the partnership initiative. The draft activity sequence plans and reflective engagement in the face-to-face professional planning conversation formed the first assessed activity, as a group assignment. The second related assessment task was an individual written reflection which required PSTs to analyse and evaluate their planning and implementation of DM Day activities. They also needed to reflect on their experiences of team teaching.

As a process of preparing, implementing and reflecting, the total time dedicated to the partnership initiative in the semester was 6 weeks.

The 'Design and Make Day' initiative arose from science teacher educators' concerns that PSTs were not having opportunities to plan and teach science and technology lessons during their regular professional experience placements (Deehan, 2022; Pressick-Kilborn & Prescott, 2017, 2020). At the same time, the partner primary school sought to enhance their science and technology education programs with a wider range of meaningful, embedded design and make tasks. A partnership with the primary school utilised a cohort approach (Lowery, 2002), and created the opportunity for carefully scaffolded success for PSTs in teaching a small group of students ahead of taking responsibility for teaching a whole class design and produce activity. The key to success was collaboration within the partnership throughout the process of planning, team teaching and reflection (Pressick-Kilborn & Prescott, 2020). Assessment tasks for the PSTs were embedded in the initiative, and recognised their creativity and adaptability in their teaching, to promote students' engagement in joyful, successful science and technology learning. As one PST reflected,

> In the same way that the children were passionately engaged by our scenarios, I really felt that this project was so exciting and captivating. It was easy to be engaged and interested. I found myself so motivated to put something exciting on for the kids. Love, love, loved it!!!. (Pressick-Kilborn & Prescott, 2017, p. 25)

Partnership initiatives such as that described in Case Snapshot 1 have impact in enhancing PSTs' confidence and enthusiasm as well as generating ideas for future teaching (Menon & Sadler, 2016) and building PCK (Kenny, 2012), all of which are carried into their careers. A description of some of the specific activity sequences that the PSTs designed and successfully implemented for the 'Design and Make Day' is included in Pressick-Kilborn and Prescott (2017), including marble runs with Kindergarten, habitats for ants or garden snails with Year 1, and Solar System tourists with Year 6. PSTs learnt about creating rich tasks to promote primary students' critical and creative thinking as well as collaboration in a meaningful and memorable science and technology school-university partnership context (Pressick-Kilborn & Prescott, 2020).

3.3.2 Case Snapshot 2: Primary PST Professional Experience Placements in Science Museums, Field Study Centres and Cultural Sites

When earlier research conducted by Griffin (2007) showed that in-service teachers often did not have the professional knowledge to maximise the learning possibilities of field trips and excursions, she worked with science education colleagues at her university and partner organisations to embed opportunities for PSTs to undertake short 'internships' with educators in science museums and field study centres (Griffin, 2012).

A primary science elective subject within a generalist Bachelor of Education degree was initially developed and offered as a pilot initiative. The elective introduced 25 PSTs to pedagogical theories relating to children's learning in a range of informal settings and created the opportunity for them to work in a small, collaborative group of three to four to complete an internship at an informal learning site. Sites included zoos, aquaria, science museums, natural history museums, and Environmental Education Centres (Buchanan et al., 2019; Griffin, 2012). During the internship, the PSTs worked alongside site educators as their mentors and observed primary students accompanied by their teachers and parent volunteers on excursions at the venue. Each small group of PSTs also engaged in a project, provided with a brief that was determined by the site educators, thus meaningful in the context of that setting. Examples of projects included creating content for museum education programs aligned with exhibits, designing education kits, and researching and suggesting possible resources for teachers to use to support students' learning before/after an excursion. Guidance was provided by the university to site educators, to ensure that the project scope suited the timeframe for the internship.

Assessment within the subject included group presentations on campus following the internship placements, which exposed all PSTs to a range of sites, projects and possibilities for student learning. The final assessment task was designed to integrate the PSTs' campus- and placement-based learning, as a reflective individual essay on ways that primary teachers can enhance students' science learning on excursions and how to work with informal learning educators.

Over time and in subsequent iterations, this concept broadened to primary PSTs developing their understanding of student learning in informal settings across a range of subject disciplines, including the creative arts, history, and geography. As a result, a third year Professional Experience subject was designed as a mandatory subject for PSTs completing a 4-year Bachelor of Education degree. On campus, the subject created opportunities to learn about how to facilitate excursions, field trips and incursions to promote student engagement and enhance learning (Griffin, 2012). The practical logistics of arranging an excursion, including risk assessment documentation and parent/carer consent, were addressed. Within the subject's delivery, students undertook a 5-day Professional Experience placement in a small group as 'apprentices' within the informal setting's learning team at diverse sites including art galleries, history museums, city farms, libraries, theatre companies, and field study centres. A requirement was that the site had to have at least one teacher employed in the education team who could formally supervise the PST interns and provide them with opportunities to focus on teaching and learning.

Locating primary PST science education in partnership activities beyond schools creates new possibilities for developing an understanding of students' learning of science. It further engages PSTs themselves in rich, meaningful, real-life settings to develop their own science knowledge and understanding, as well as possibilities for experiencing awe (Cooke-Nieves et al., 2022). Lemon and Weller (2015) claim that such partnerships with cultural organisations have been underutilised by teacher education, including more specifically in primary science teacher education. As Griffin (2012) notes, "Teacher education programs concentrate almost exclusively on classroom environments, and yet much learning by school students involves experiences beyond the classroom, in informal learning settings of many kinds" (p. 117). The partnerships established between universities and a range of cultural community sites can afford unique opportunities for PST professional learning experiences as well as providing insight into alternative future careers in education. Furthermore, Case Snapshot 2 highlights the potential role of educators in other settings to contribute to primary PST science education.

3.3.3 Case Snapshot 3: Local Primary School Students Visiting the University Campus for a PST-led Science-Focused Workshop

In this initial core, mandatory primary Science and Technology Education subject, the PSTs (86) worked in small teams of two to four, to engage in the process of co-teaching a 90-min standalone, hands-on science workshop to small groups of six to eight Year 2 or Year 4 students (147 students in total). The students attended a nearby primary school within walking distance of the campus, and their teachers volunteered for their classes to participate as part of a wider ongoing school-university partnership arrangement.

Prior to the workshop, an optional visit to the school was arranged outside of the on-campus tutorial schedule, for PSTs to observe the students learning Science and Technology in their regular classroom, which some but not all PSTs took up. During the on-campus tutorials leading up to the workshop, the PSTs formed teaching teams and received guidance and support from their tutors to design a lesson that included:

(1) *an initial stimulus to focus and engage or hook the students,*
(2) *a brief informal assessment of students' relevant prior knowledge,*
(3) *an introduction to the hands-on task, which could include explicit teaching,*

(4) *completion of the task with differentiated 'just-in-time' support from the PSTs, and*

(5) *culmination of the workshop (including sharing products of learning and reflection on learning).*

The PST teaching teams wrote a detailed plan for their workshop that included specific learning outcomes and lesson timing. They identified and noted specific roles for each PST in the lesson delivery and prepared resources necessary for students to complete the task. They developed their own knowledge and understanding of the concepts and content of the lesson, so that they felt well-prepared to teach. On the day of the workshop, the primary students visited campus with their class teachers[3] during the regular tutorial time for the PSTs. Students participated in a workshop led by a PST teaching team (see Fig. 3.2)*, supervised by the tutors and their class teachers.*

Following the workshop delivery, the PSTs engaged in reflective discussion with question prompts provided by the tutor to guide evaluation. An assessment task in two parts was associated; first, a small group task in which the lesson plan was submitted and feedback from the tutor provided to PSTs prior to the workshop delivery. This timing allowed for adjustments to be made to the workshop plans before implementation. Secondly, an individual written reflection was submitted for assessment following the teaching of the workshop.

Often in PST education, assessment tasks are devised which involve planning a lesson or a sequence of lessons, to teach a particular concept to specified-age primary students, designed to meet certain Syllabus learning outcomes. The purpose of such a task lies in familiarising PSTs with the planning process, including identification of relevant Syllabus outcomes, articulation of learning intentions and success criteria, alignment of teaching and learning activities with the intended learning outcomes and the purposeful selection of resources. This can be a valuable professional learning activity in and of itself. The problem with such assessment tasks, however, is that the PSTs do not usually have an opportunity to actually teach the lesson or lesson sequence planned to primary students; it remains a hypothetical, abstract or ideal lesson. It is in the *teaching* of the lesson, with support from more expert educators, and the subsequent reflective evaluation on the lesson, that rich learning about primary science pedagogy becomes possible and PST self-efficacy is enhanced (Mansfield & Woods-McConney, 2012). School-university partnerships, such as the one in this case snapshot, can create shared contexts for co-designed lessons to be delivered with specific focus and support, outside of the traditional professional experience placement.

In Case Snapshot 3, inviting primary students from a local partner school to the university campus to participate in hands-on, standalone workshops resulted in the PSTs being able to put into action the lessons that they had collaboratively designed. They received support from tutors, peers and in-service teachers from the local school

---

[3] The school's staff arranged the visit as an excursion with the necessary risk assessment and consent from parents.

Fig. 3.2 Collage of workshop photos

to plan, implement and reflect on the lesson taught to a small group of primary students. The PSTs had opportunities in the moment to make responsive decisions, to make the tasks more engaging, change the pace, or capitalise on students' interest. By being adaptable and responsive in their lesson delivery, PSTs gained confidence in their ability to teach a successful science lesson that incorporated their own scientific understanding and specific science pedagogies that they were learning about in their generalist primary ITE degree.

3.3.4   Other Australian School-University Partnerships in Primary PST Science Education

As Ma and Green (2023) note, school-university partnerships are not uncommon in Australia for strengthening teacher education reform. An Australian multi-university project, Science Teacher Education Partnerships with Schools (STEPS), provides other rich examples of science-focused partnerships to promote primary PST education (see Jones et al., 2016 for case descriptions; vignettes and full case studies can be accessed from http://www.stepsproject.org.au). A variety of partnership models with different-sized PST cohorts and timeframes, as well as at different timepoints within ITE degrees, are featured in the case studies. In some models, the in-service teachers were involved and in other models, the teacher educator was present at the school, but not in all cases. In some case studies, the associated tutorials or workshops were held at the school rather than on-campus. In all cases, an assessed task for the PSTs was incorporated into the partnership model. Taken together, these case studies provide inspiration for alternative modes of delivering primary science PST education across regional and city university contexts.

3.4   Looking to the Future: What Role can Partnerships Further Play in Preparing 'Classroom-Ready' Graduate Teachers?

'Classroom-ready' graduate primary science teachers need joyful, diverse and discipline-specific professional experiences during their ITE degree, which partnership initiatives create. Such first-hand experiences prepare graduates to capably and creatively plan for and support primary students in their science learning, with insight into how to design engaging lessons using contemporary approaches in diverse contexts. While the triumphs of partnerships have been shared through the case snapshots in this chapter, some of the tensions and challenges in building and maintaining partnerships also warrant discussion. Such challenges include the changing nature of relationships, roles and responsibilities, and the time that it takes to grow and sustain a partnership.

In each of the case snapshots described and analysed in the previous section, the contribution of the partnership activities to the PSTs' professional learning created unique opportunities to see and teach science. In particular, the supported cohort approach with a selected partner school or institution ensured the quality of the experience, as partners were identified as competently designing and implementing contemporary primary science programs. Each partnership highlighted in the case snapshots came to an end largely as a result of ITE degree redesign or changes in staff coordination of subjects as well as changes in education leadership in the partner schools and organisations. Partnerships ultimately rest on developing and maintaining positive interpersonal relationships (Jones et al., 2016) and rely on the good intentions of staff in both schools and universities to take action, respectful of the diverse contributions that all stakeholders can bring. A challenge lies in sustaining school-university connections when they are built solely on relationships between individuals, however, rather than more formalised institutional arrangements (Deehan, 2022).

While teaching may be 'a profession based on partnership' (European Commission, 2005 cited by Farrell, 2023), there can be ambiguity around roles and responsibilities, bringing many challenges to collaboration with numerous levels of micro politics (Farrell, 2023). In seeking to represent partnership practice, including the roles and levels of responsibility, Jones et al. (2016) have developed a typology of school-university partnerships, defined by purpose, nature and levels of embeddedness. The three levels of partnership embeddedness are labelled as (i) Connective, (ii) Generative, and (iii) Transformative, each with distinct value in meeting specific purposes and outcomes (Jones et al., 2016). Connective partnerships are characterised by one partner having a particular need that can be serviced by the other, but with benefit and value to all stakeholders. Such partnerships are often short-term in nature, and Case Snapshot 3 could be considered as an example, although this one-off PST-led workshop took place in the context of a wider partnership between the university and school. Generative partnerships result in new or different practices or structures emerging as a result of responsiveness to partners' needs. Case Snapshot 1 could be considered as an example of a Generative partnership, as the DM Day was developed in response to the need to create a science- and technology-specific professional experience for PSTs (university partner), and to embed design and produce tasks in an existing science program (school partner). Finally, Transformative partnerships "are on-going and embedded in the programs of the collaborating institutions" (Jones et al., 2016, p. 115), with the learning through the partnership resulting in transformation of practices by either the school or university. Case Snapshot 2 is an example of such a partnership, as all partner members were engaged in the planning and delivery of the PSTs' experiences in the informal science learning contexts of museums and other sites. At both the university and informal sites, critical reflection resulted in various iterations of the collaboration over time.

In primary ITE degrees, the pressure of time within programs for meaningful and impactful science teacher education, and in forming and sustaining partnerships more specifically, has been frequently raised as an issue within the literature (e.g., Deehan, 2022; Palmer, 2015), as well as in the case snapshots shared here.

The passage of time necessary for the evolution of partnerships—establishment, development, maintenance, extension—is acknowledged by Ma and Green (2023), with the nature of the partnership and the relationships and roles within it changing over time (Jones et al., 2016). Furthermore, there is concern expressed about the time necessary to allow for greater involvement of in-service teachers and informal site educators in negotiating, designing, and directing ITE partnership initiatives to ensure trust, mutuality and reciprocity (Gilbert & Hobbs, 2022), otherwise, the partnership agenda may privilege university needs (Ma & Green, 2023). Ma and Green (2023) also caution against school-university partnership initiatives that create additional work for in-service teachers, emphasising the importance of embedding activities in existing work, towards meeting the expectations of the partner school. Building 'collaborative intention' or collective vision and identifying challenges then working towards a solution in subsequent iterations, are considered key to the success of cross-institutional partnerships (Ma & Green, 2023).

Despite these potential barriers and obstacles, we maintain that forming and sustaining partnerships remains a promising endeavour, integral to the future of primary science teacher education in Australia and internationally. In this post COVID-19 lockdown era, there is perhaps an even greater importance of in-person, first-hand immersive experiences in primary schools as an embedded feature of PSTs' professional learning. This is amplified in cases of online delivery of primary science education subjects—PSTs need to see and experience contemporary science teaching in real classrooms that extends beyond the traditional practicum placement (Neal & Eckersley, 2014). From the university partner's perspective, renewal of ITE degrees creates opportunities for embedding partnership initiatives (Ma & Green, 2023), while reform and revisions to the science curriculum by governing bodies such as ACARA (Australian Curriculum, Assessment and Reporting Authority)[4] can provide impetus to the formation or repurposing of partnerships for all stakeholders. In doing so, partnerships can create unique opportunities to richly enhance the quality of primary science teacher education, and as a result, positively impact the preparation of 'classroom-ready' graduate teachers.

References

Australian Academy of Science. (n.d.). *Primary connections: Linking science with literacy.* Accessed May 9, 2023. https://primaryconnections.org.au/

Buchanan, J., Pressick-Kilborn, K., & Maher, D. (2019). Promoting environmental education for primary school-aged students using digital technologies. *Eurasia Journal of Mathematics, Science and Technology Education, 15*(2). https://doi.org/10.29333/ejmste/100639

Bybee, R. W. (1997). *Achieving scientific literacy: From purposes to practices.* Heinemann.

Cooke-Nieves, N., Wallace, J., Gupta, P., & Howes, E. (2022). The magic of informal settings: A literature review of partnerships and collaborations that support preservice science teacher

---

[4] More information about ACARA and Australia's National Curriculum can be accessed here https://www.australiancurriculum.edu.au/.

education across the globe. In J. A. Luft & M. G. Jones (Eds.), *Handbook of research on science teacher education* (pp. 189–202). Routledge.

Craven, G., Beswick, K., Fleming, J., Fletcher, T., Green, M., Jensen, B., Leinonen, E., Rickards, F. (2014). *Action now: Classroom ready teachers.* Teacher Education Ministerial Advisory Group. Accessed April 12, 2023 at https://www.aitsl.edu.au/tools-resources/resource/action-now-classr oom-ready-teachers

Daza, V., Gudmundsdottir, G. B., & Lund, A. (2021). Partnerships as third spaces for professional practice in initial teacher education: A scoping review. *Teaching and Teacher Education, 102*, 1–14.

Deehan, J. (2022). Primary science education in Australian universities: An overview of context and practice. *Research in Science Education, 52*(6), 1735–1759. https://doi.org/10.1007/s11 165-021-10026-6

Farrell, R. (2023). The school-university nexus and degrees of partnership in initial teacher education. *Irish Educational Studies, 42*(1), 21–38. https://doi.org/10.1080/03323315.2021.189 9031

Fitzgerald, A, Pressick-Kilborn, K., & Mills, R. (2021). Primary teacher educators' practices in and perspectives on inquiry-based science education: Insights into the Australian landscape. *Education 3–13, 49*(3), 344–356.

Forbes, C. (2013). Curriculum-dependent and curriculum-independent factors in preservice elementary teachers' adaptation of science curriculum materials for inquiry-based science. *Journal of Science Teacher Education, 24*, 179–197. https://doi.org/10.1007/s10972-011-9245-0

Gilbert, A., & Hobbs, L. (2022). Partnerships in K-12 preservice science teacher education. In J. A. Luft & M. G. Jones (Eds.), *Handbook of research on science teacher education* (pp. 178–188). Routledge.

Griffin, J. (2007). Students, teachers and museums: Toward an intertwined learning circle. In J. H. Falk, L. D. Dierking, & S. Foutz (Eds.), *In principle, in practice: Museums as learning institutions* (pp. 31–42). Altamira Press.

Griffin, J. (2012). Exploring and scaffolding learning interactions between teachers, students and museum educators. In E. Davidsson & A. Jakobsson (Eds.), *Understanding interactions at science centers and museums: Approaching sociocultural perspectives* (pp. 115–128). Brill.

Jones, M. (2008). Collaborative partnerships: A model of science teacher education and professional development. *Australian Journal of Teacher Education, 33*(3), 61–76.

Jones, M., Hobbs, L., Kenny, J., Campbell, C., Chittleborough, G., Gilbert, A., Herbert, S., & Redman, C. (2016). Successful university-school partnerships: An interpretive framework to inform partnership practice. *Teaching and Teacher Education, 60*, 108–120. https://doi.org/10. 1016/j.tate.2016.08.006

Jones, M. G., & Carter, G. (2007). Science teacher attitudes and beliefs. In S. K. Abell & N. G. Lederman (Eds.), *Handbook of research on science education.* Routledge.

Kenny, J. D. (2012). University-school partnerships: Pre-service and in-service teachers working together to teach primary science. *Australian Journal of Teacher Education, 37*(3), 57–82.

Kenny, J. D., Hobbs, L., Herbert, S., Chittleborough, G., Campbell, C., Jones, M., Gilbert, A., & Redman, C. (2014). Science teacher education partnerships with schools (STEPS): Partnerships in science teacher education. *Australian Journal of Teacher Education, 39*(12), 43–65.

Lemon, N., & Weller, J. (2015). Partnerships with cultural organisations: A case for partnerships developed by teacher educators for teacher education. *Australian Journal of Teacher Education, 40*(12). https://doi.org/10.14221/ajte.2015v40n12.4

Lowery, N. V. (2002). Construction of teacher knowledge in context: Preparing elementary teachers to teach mathematics and science. *School Science and Mathematics, 102*(2), 68–83.

Ma, H., & Green, M. (2023). A longitudinal study on a place-based school-university partnership: Listening to the voices of in-service teachers. *Teaching and Teacher Education, 129*. https://doi. org/10.1016/j.tate.2023.104148

Magee, P. A., & Flessner, R. (2012). Collaborating to improve inquiry-based teaching in elementary science and mathematics methods courses. *Science Education International, 23*(4), 353–365.

Mansfield, C. F., & Woods-McConney, A. (2012). I didn't always perceive myself as a science person: Examining efficacy for primary science teaching. *Australian Journal of Teacher Education, 37*(10), 37–52.

Menon, D., & Sadler, T. D. (2016). Preservice elementary teachers' science self-efficacy beliefs and science content knowledge. *Journal of Science Teacher Education, 27*(6), 649–673.

Neal, G., & Eckersley, B. (2014). Immersing pre-service teachers in site-based teacher school-university partnerships. In M. Jones & J. Ryan (Eds.), *Successful teacher education: Partnerships, reflective practice and the place of technology* (pp. 31–48). Sense Publishers.

New South Wales Department of Education. (2021). *Co-teaching: A handbook of evidence for educators* (2nd ed.). State of NSW (Department of Education). Accessed May 9, 2023. https://education.nsw.gov.au/content/dam/main-education/en/home/teaching-and-learning/school-learning-environments-and-change/future-focused-learning-and-teaching/Co_teaching_handbook_2nd_edition.pdf

New South Wales Education Standards Authority (NESA). (2017). *Science and technology K-6 syllabus.* Accessed May 9, 2023. https://educationstandards.nsw.edu.au/wps/portal/nesa/k-10/learning-areas/tas/science-and-technology-k-6-new-syllabus

Palmer, D. (2015). Maintaining the balance: Creative practices in university-school partnerships for teacher education. *Creative Education, 6*, 1530–1535.

Petersen, J. E., & Treagust, D. F. (2014). School-university partnerships: The role of teacher education institutions and primary schools in the development of preservice teachers' science teaching efficacy. *Australian Journal of Teacher Education, 39*(9), 153–167.

Pressick-Kilborn, K., & Prescott, A. (2017). Engaging primary children and pre-service teachers in a whole school 'design and make day': The evaluation of a creative science and technology collaboration. *Teaching Science, 63*(1), 18–26.

Pressick-Kilborn, K., & Prescott, A. (2020). School-university partnerships as rich STEM learning contexts for preservice teachers working with primary students. In A. Fitzgerald, C. Haeusler, & L. Pfeiffer (Eds.), *STEM education in primary classrooms: Unravelling contemporary approaches in Australia and New Zealand* (pp. 100–114). Routledge.

Zeichner, K. (2010). Rethinking connections between campus courses and field experiences in college- and university-based teacher education. *Journal of Teacher Education, 61*(1&2), 89–99.

Chapter 4
How Are Science Specialisations Conceptualised and Realised in Primary Teacher Education?

I have more questions than answers about primary science
specialisations.
Reece

4.1 Introduction

This chapter takes a deep dive into a recent initiative in science teacher education already mentioned in Chaps. 1 and 2—primary specialism. Chapter 1 identified the training/hiring of science/STEM specialists as an area of national priority in an effort to raise the volume and quality of science teaching in schools. Chapter 2 briefly problematised generalist versus specialist primary science teachers, arguing for the prioritisation of science by school leaders regardless of the method of delivery. This chapter hones in on primary specialisation as it relates to initial teacher education (ITE) policy—these educators are distinct from specialist teachers. Recent reforms in ITE in Australia require pre-service teachers to graduate with a specialisation in a priority learning area that may be science alongside their regular study. This is a significant change to the way that primary teachers are prepared, having ramifications for the content and structure of ITE and school science education. This chapter maps the what, who, why, and how of primary science specialisation through one author's research program, answering the questions: What is a teacher with a science specialisation?; Who is a teacher with a science specialisation?; Why are teachers with a science specialisation needed (or not)?; and How do teachers with a science specialisation work successfully in schools?

This chapter is divided into five parts. First, we outline a brief literature around teacher specialism. Second, we describe the origins of primary specialisation in Australian ITE. Third, we draw on one author's research program to generate a 'research assemblage' that lays out empirical research conducted on the topic. The research assemblage comprises five empirical research papers, which are analysed to answer the above questions. Fourth, we present findings about the what, who, why, and how of primary science specialisation. Finally, we consider triumphs and

A. Fitzgerald et al., *Contemporary Australian Primary Science Teacher Education*, SpringerBriefs in Education, https://doi.org/10.1007/978-981-97-5660-5_4

tensions related to primary science specialisation and give future recommendations for science teacher education.

4.2 A Brief Literature

The Cambridge dictionary defines 'specialisation' as "the process of concentrating on and becoming expert in a particular subject or skill" (Cambridge Dictionary, 2022a). While this definition is seemingly straightforward, commentary tends to use variations on this term such as specialist teacher, enhancement teacher, or mentor teacher. Scholarship also extends to the use of the term instructional coaches. In this brief review of global research, we have used the term specialism to be inclusive of these various educators as well as purposefully used terminology offered by Mills et al. (2020) in their tripartite model of teacher specialism. This model distinguishes between instructional coaches, specialist teachers, and teachers with a specialisation (both in-service and pre-service). Their role and responsibilities, as well as (desired) results, are shown in Fig. 4.1.

For the most part, the global literature appears to be located in the USA where specialist teachers and instructional coaches have been encouraged through policy and professional associations and industry organisations. This is especially true in maths and science education. In the US there are strong calls from the National Council of Teachers of Mathematics (NCTM) for the use of specialist teachers to enhance the teaching, learning and assessing of mathematics and to improve student learning. Similarly, the National Science Teachers Association (NSTA) promotes science-specific teaching support for teachers in the form of science mentors within and across grade levels (NSTA, 2017, 2018). In the UK, there are calls for specialist science teachers, underpinned by the argument that strong subject knowledge impacts the effectiveness of science teaching (Association for Science Education [ASE], 2019). A range of nine industry organisations recently lobbied to the ASE in support of subject-specific mentoring in primary schools (ASE, 2019).

In Australia, some states have successful primary science and mathematics specialist teacher programs (e.g., Victoria's Primary Mathematics and Science Specialists Initiative; Department of Education and Training [DET], 2019), but most do not, and some have been quite unsuccessful (e.g., Queensland's Primary Science Facilitators; Department of Education, Training and Employment [DETE], 2012). There is very recent interest in engineering and STEM specialist teachers. In Queensland, for instance, in 2019–21 the Department of Education and Training committed to ensuring every state school had access to a Science, Technology, Engineering and Mathematics (STEM) specialist teacher or 'STEM Champion' as part of an $81.3 million STEM education package. Despite massive investments from governments nationally and internationally, there is surprisingly little scholarship about primary specialisms.

There is an emerging scholarship about the teaching and leading practices of primary school teachers with a science specialism. Mills et al. (2020) have conducted

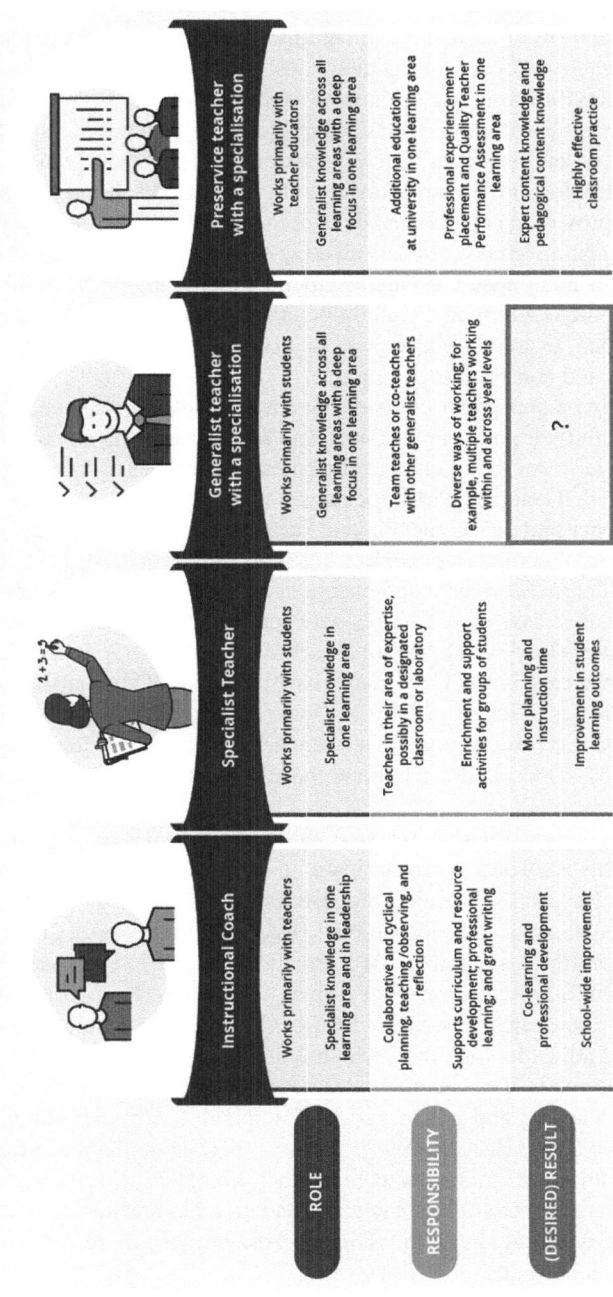

Fig. 4.1 Tripartite model of teacher specialism (adapted from Mills et al., 2020)

a systematic review of the global academic literature. These authors elaborate tasks typically carried out by teachers with a science specialism may include collaborative curriculum and lesson planning, including locating resources for teachers; observing a colleague's science or mathematics teaching; modelling science or mathematics teaching; co-teaching science or mathematics lessons; working with students; pre- and post-lesson conferences and reflective conversations; and handling student assessment or learning data. To a lesser extent, and particularly in the case of instructional coaches, they might be involved in facilitating and sharing information amongst teachers; providing professional development; and writing or administering grants. Other administrative tasks such as meeting with school leadership teams and responding to emails and phone calls are also part of their daily work. While seemingly well defined, the teaching and leading practices of teachers with a specialism require further critical examination, with a paucity of scholarship looking at what factors may enable and constrain their practice.

Other scholarship on subject specialisms outside of teachers' roles and responsibilities has dealt with their personal attributes and effectiveness on student achievement in science. In terms of their personal attributes, two articles in the international literature suggested that teachers with a specialism were more likely to have a formal qualification in science and thus higher content knowledge (Brobst et al., 2017; Nickerson, 2010). However, neither article linked this with student achievement. The articles concerning student achievement demonstrated conflicting findings, most showed no difference in science assessment scores between students taught by a generalist and specialist teacher (Levy et al., 2016; Roach & Wendt, 2022; Schwartz et al., 2000) and one showed a greater mathematical achievement of students taught by a specialist maths teacher (Nickerson, 2010). Most recently a US study of 282 fifth-grade students found that no difference exists between the science achievement scores of students taught by a specialist teacher and those who were not (Roach & Wendt, 2022).

The broader educational impact of teachers with a specialism is also reported in the global literature. This is related to greater instruction time and professional learning and collaboration. Research generally suggests more time is afforded to science instruction in a specialist/specialisation model (Brobst & Markworth, 2019; Brobst et al., 2017; Markworth et al., 2016, 2018; Poland et al., 2017), perhaps because of the time allocated to science by way of formal timetabling. However, a study by Levy et al. (2016) found such scheduling restricted instructional time because of a lack of flexibility to adjust lesson times, which is afforded in a generalist classroom. Common to studies focused on instructional coaches was their benefit to school-wide professional learning and collaboration (Campbell & Chittleborough, 2014; Dailey & Robinson, 2016; Harvey, 1999), whereas there is no evidence that other models of specialism afford this advantage. In fact, Webel et al. (2017) found the specialist teacher in their study felt isolated and had no other teachers to collaborate with (see also Markworth et al., 2016). This is likely because there was no shared responsibility for science education (Levy et al., 2016).

4.3 Primary Specialisation in Australian Initial Teacher Education

Primary specialisation in Australian education originated in the Teacher Education Ministerial Advisory Group's (TEMAG) final report, *Action Now: Classroom Ready Teachers*, in late 2014. In response to "the challenges primary teachers face in … delivering instruction across the … range of subject areas in the primary curriculum", the Advisory Group's Recommendation 18 stated "Higher education providers equip all pre-service teachers with at least one subject specialisation, prioritising science, mathematics or a language" (TEMAG, 2014, pp. 21–22). The Government response, in early 2015, endorsed the report's recommendations, using national course accreditation requirements to "make sure that every new primary teacher graduates with a subject specialisation" (DET, 2015, p. 8). In 2015 (updated 2022), the Australian Institute for Teaching and School Leadership (AITSL) defined a teacher with primary specialisation in Program Standard 4.4 as possessing "expert content knowledge and pedagogical content knowledge and highly effective classroom teaching" in their area of specialisation (AITSL, 2022). Accreditation guidelines clarified in 2017 (updated 2020) this was in addition to study in each of the learning areas of the primary school curriculum:

> Graduates will be prepared to teach in each of the learning areas of the primary school curriculum, and *in addition*, will undertake a specialisation in one learning area of the primary school curriculum. Thus, primary graduates of initial teacher education (ITE) programs will be generalist primary teachers, with a deep focus in a particular learning area. (AITSL, 2020, p. 2, emphasis added)

A stimulus paper was released by AITSL, which elaborated primary specialisation by providing examples of their roles and responsibilities (AITSL, 2017). In brief, their suggestions include confidence in and enthusiasm for science teaching, a deep understanding of science concepts and processes, and a deep understanding of scientific inquiry.

Models of primary specialisation in teacher education programs have only been running for a few years so there is very little scholarship in this area. Main et al. (2021) have conducted a related audit of teacher education programs in Australia from the information made available in programme handbooks and on university websites. These authors found that the most common subjects offered in primary specialisation were mathematics/numeracy and English/literacy, followed by science. Other areas of specialisation not offered by many universities include humanities and social science (HASS), the Arts, inclusive education, and Catholic education. There appear to be very diverse ways of approaching primary specialisations. Some universities offered several units in their program over to specialisations (ranging from 1 to 5 units or subjects) including subjects run by education faculties or other areas of the university like science faculties. There remains a paucity of research describing and evaluating practice in this area, so gaining a complete picture of the national landscape is challenging. Main and her colleagues (2021) suggest the variance in provision offered by Australian universities highlights potential obstacles which need to be considered to ensure quality across all programmes—the main objective of AITSL's program standards.

4.4 Research Assemblage

The term assemblage is defined as a product that is made of different things put together for a particular reason (Cambridge Dictionary, 2022b). The assemblage in this chapter is made of empirical research articles with different research foci related to primary science specialisation, put together for the purpose of a meta-analysis. The research assemblage includes a systematic literature review and policy analysis as well as articles with diverse research participants representative of critical stakeholders: school teachers, pre-service teachers, and teacher educators. These papers comprise the most substantive line of inquiry on primary specialisation in Australian ITE—a relatively new practice and scholarship.

Following Braun and Clarke (2006), the papers were analysed to produce a thematic map—"an overall conceptualisation of the data patterns, and relationships between them" (p. 89). This was done by reading and re-reading the papers and drawing out empirical findings as they relate to the central what, who, why, and how questions guiding this chapter: What is a teacher with a science specialisation?; Who is a teacher with a science specialisation?; Why are teachers with a science specialisation needed (or not)?; and How do teachers with a specialisation work successfully in schools? The findings were coded and eventually the codes were collapsed into themes. The final thematic map, presented in the next section, lays out the what, who, why, and how of primary science specialisation in Australian education. Each paper in the research assemblage is now briefly summarised before the findings and discussion are presented.

Mills, R., Bourke, T., & Siostrom, E. (2020). Complexity and contradiction: Disciplinary expert teachers in primary science and mathematics education. *Teaching and Teacher Education, 89,* **1–12.**

This literature review aimed to determine what is known about "disciplinary expert teachers"—an umbrella term that includes instructional coaches, specialist teachers, or teachers with a specialisation—in primary science and mathematics education. Following systematic literature review protocols, 37 articles were included in the analysis. Critical amongst the findings was the highly varied roles and responsibilities of these educators.

Bourke, T., Mills, R., & Siostrom, E. (2020). Origins of primary specialisation in Australian education: What's the problem represented to be? *Australian Educational Researcher, 47,* **725–740.**

In this paper, Carol Bacchi's What's the Problem Represented to be? approach (WPR) was used to analyse the discourses present in four policy documents related to primary specialisations. The findings revealed that primary teachers are positioned as lacking specialised content knowledge in science, mathematics/numeracy, and English/literacy. The authors outline how mechanisms such as repetition and co-location of statements as well as "cherry-picked" research and statistics culminates in a problem representation that ought to be thought about differently.

Mills, R., & Bourke, T. (2020). Primary specialisation in Australian education: Pre-service teachers' lived experiences. In J. Fox, C. Alexander, & T. Aspland (Eds.), *Teacher education in globalised times* **(pp. 23–40). Springer.**

This narrative inquiry gave voice to five primary pre-service teachers completing a science specialisation at an Australian university. Analysis of interview data suggested a balance of opportunities and challenges related to primary specialisations. Pre-service teachers found online modules relevant, valuable, and transformative, but concomitantly time consuming, confusing, and difficult.

Bourke, T., & Mills, R. (2022). How teacher educators do policy: Enacting primary specialisations. In T. Bourke, D. Henderson, R. Spooner-Lane, & S. White (Eds.), *Reconstructing the work of teacher educators* **(pp. 31–50). Springer.**

In this chapter, the reflexive accounts of two teacher educators illuminate the operationalisation of a primary science specialisation at an Australian university. The work of these two 'policy actors' showed that despite being constrained by various factors such as course accreditation, space for agentic ways of working opened up as policy interpretations moved from one actor to the next.

Mills, R., & Bourke, T. (in preparation). What practice architectures enable and/or constrain primary school teachers with a science/STEM specialism?

This qualitative inquiry, underpinned by the theory of practice architectures, draws upon data generated from interviews with five primary school teachers who identified as having a science/STEM specialism. Cultural-discursive (sayings), material-economic (doings), and socio-political (relatings) arrangements were analysed to answer the question: What practice architectures enable and/or constrain primary school teachers with a science/STEM specialism? Findings revealed the importance of a supportive school culture and professional learning networks as "nested practice architectures" (practice architecture that enable/contain other practice architectures).

4.5 Findings and Discussion

Four themes were generated from the research assemblage (Table 4.1). These themes are now elaborated to highlight how binaries in the field present both tensions and triumphs for educators and students at both school and university levels.

Table 4.1 Themes generated from the research assemblage

What?	Theme 1: generalist versus specialist teacher
Who?	Theme 2: novice versus expert teacher
Why?	Theme 3: student enrolment and engagement versus achievement
How?	Theme 4: high versus low status in schools

4.5.1 Theme 1: Generalist Versus Specialist Teacher (What)

Theme 1 answers the question: What is a teacher with a primary specialisation? A teacher with a primary specialisation is defined across multiple policy documents and a handful of research studies. Despite frequent slippage between the terms "specialist" and "specialisation", there is now a relatively clear picture of what is a teacher with a specialisation. A teacher with a specialisation is a generalist teacher who predominantly works with students. These educators have generalist knowledge across all learning areas with a deep focus in one learning area, for example science. Not much is known about these educators' practices because there is a lack of international research documenting this approach. It appears that individual schools define the role and responsibilities of these educators, sometimes giving them titles such as "science co-teacher" and "STEM teacher" (Mills & Bourke, in preparation). These educators may team teach or co-teach with other teachers (i.e., share classes), working both within and across year levels.

The specialisation approach causes tension because it positions pre-service teachers as both "specialist" and "generalist". The positioning of pre-service teachers with a specialisation as specialist is evident in ITE accreditation policies by the repetition of words such as "expert" and "expertise", often co-located with terms such as "knowledge" (Bourke et al., 2020). This is coupled with the articulation that a teacher with a primary specialisation will "assist other teachers … to teach the subject effectively" (AITSL, 2020, p. 2). This suggests that leadership and professional learning are responsibilities of a teacher with a specialisation. Concomitantly, however, these policies specify specialisation in addition to or on top of their regular duties as a generalist classroom teacher: "This does not mean primary teachers will teach only in their area of specialisation" (AITSL, 2020, p. 2). According to the professional accreditation guidelines document, authored by AITSL, teachers with a specialisation are defined as "graduates who are *generalist primary teachers* with a specialisation. It is important that these graduates are identified as *distinct from specialist teachers* who fulfil specialist roles in schools such as librarian or health/physical education" (AITSL, 2020, p. 3, emphasis added). As Bourke et al. (2020) rightfully question: "If pre-service teachers are to be the 'specialists', then how can they simultaneously be generalists?" (p. 732). This has led to confusion around enacting this policy.

Stakeholders such as pre-service teachers and teacher educators echoed the sentiment that primary specialisations are confusing. Drawing from Mills and Bourke's (2020) study with pre-service teachers, one participant admitted she did not know a lot about the science specialisation—"I don't know a lot about it, sorry"—but described how this might "take the workload off generalist teachers" (p. 33). This suggests she believed she might be solely responsible for teaching science within/ across year levels. Another pre-service teacher specified that for her, "[science] will be my area that I am responsible for having expertise in" (p. 33). She elaborated on what this meant by giving examples of the types of activities she might be involved with, which included "reviewing what's already there [science education broadly]", "planning and sequencing of lessons", and "a role in the professional development of that particular area for teachers" (p. 33). This pre-service teacher thought she would have a leadership role to play in science education at her school. Yet another pre-service teacher thought of the science specialisation as a means to "become a better science teacher" and "lead by example" (p. 33). Taken together, these data suggest pre-service teachers know very little about primary specialisations and have differing views about the role of a teacher with a specialisation (cf. Main et al., 2021).

Defining what counts as a specialisation also presented confusion for teacher educators who had to enact the primary specialisation policy (Bourke & Mills, 2022). For two teacher educators enacting primary specialisations at an Australian university, interpreting the relevant policies and translating them into practice was challenging. Both teacher educators grappled with what it meant to have a specialisation that was *additional* to what is already provided in the teacher education program. Reflecting upon their enactment of the policy, these teacher educators managed to embed an additional level of content and pedagogical knowledge, however identified leadership knowledge and skills as an area that was lacking. This seems to be the case internationally when looking at how to prepare specialist teachers or teachers with a specialisation—there is very little research evidence that documents the upskilling in leadership needed for these roles and how this is achieved (cf. Mills et al., 2020).

What becomes clear when turning to look from ITE policy to school practice is that hybrid specialist-generalist ways of working are already happening in Australian schools, but also with great tension. The lead author of this chapter is located in Queensland, where primary teachers are prepared as generalists to teach across all learning areas in the national curriculum. There are, however, a mix of formal and informal science specialism practices operating across state, independent, and Catholic school sectors. With regard to the largest state sector, the Queensland Department of Education has trialled numerous initiatives related to teacher specialism with limited ongoing funding and success. In 2010–12, an initiative titled Science Spark was enacted wherein $37.7 million was committed over three years to the employment of 15 'Regional Managers (Science)' and 100 'Primary Science Facilitators' to overhaul teaching and learning in science across Queensland. These educators travelled across the state delivering professional development to primary school teachers as well as worked in schools to support teachers through collaborative planning and teaching. While interim and final reports on the success or otherwise of Science Spark are not publicly available, commentators have suggested that the Primary Science Facilitators ended up working in isolation as specialist science teachers such that "primary school teachers were no better able to teach science than previously" (Pezaro, 2017, para. 16). So, while the Science Spark program meant that there were qualified science teachers in primary schools across the state, its wider and ongoing impact is hard to determine.

Following this in 2019–21, the Queensland Department of Education committed to ensuring every state school had access to a Science, Technology, Engineering, and Mathematics (STEM) specialist teacher or 'STEM Champion' as part of an $81.3 million STEM education package. The efficacy of this second approach is yet to be seen, with funding presently ending and seemingly no research on the horizon; however, a critical challenge to science education has been the use of STEM Champions for specific STEM disciplines or skills such as digital technology or coding and robotics. In a similar manner to Science Spark, it is a wonder whether school teachers (and students) are any better off than previously in terms of science education specifically. Other approaches to specialism currently occurring in primary schools include principals appointing science specialists and teachers sharing classes of students to teach in their area/s of strength or interest that may include science, with the latter being mostly informal arrangements amongst teachers themselves making it hard to discern how widespread such practices are and what their impact may be (Mills et al., 2020).

The roles and responsibilities as well as desired outcomes of a teacher with a specialisation is closely linked to who takes on these roles. This is the focus of Theme 2, which looks at who is a teacher with a specialisation and who can/should have specialised science and science teaching knowledge.

4.5.2   Theme 2: Novice Versus Expert Teacher (Who)

Theme 2 answers the question who is a teacher with a specialisation and who can/ should be a teacher with a science specialisation? Throughout accreditation policy we see the positioning of teachers with a specialisation in the role of an instructional coach or specialist. As previously outlined, teachers with a specialisation "assist other teachers … to teach the subject effectively" (AITSL, 2020, p. 2). Given that primary specialisations are for undergraduate pre-service teachers, their capacity to do this work as novice or beginning teachers is questionable. This tension is felt by pre-service teachers who struggle to reconcile being a novice (beginning teacher) and an expert (specialisation).

In Mills and Bourke (2020), one pre-service teacher noted "the term [specialisation] is very sticky for me" (p. 35). She described "coming in as a person who wants to learn yet being an expert in something seems really confused" (p. 35). Moreover, pre-service teachers thinking about the prospect of upskilling more experienced teachers describe their feelings as "nerve-racking" and "terrifying" (p. 35). Pre-service teachers questioned whether they would have the capability to impact another teacher's practice and were anxious about how more experienced teachers might feel about this. This sentiment is encapsulated by a pre-service teacher who commented they felt "insecure" because "the last thing they [in-service teachers] want to hear from a newbie is how to do it [teach science]" (p. 35). This pre-service teacher commented "I don't imagine it's going to go down well" (p. 35). There were some exceptions in Mills and Bourke's (2020) study. In opposition to this perspective, one pre-service teacher described, "I think having that extra support … would be a good thing for them [schools]" (p. 35). Similarly, a different pre-service teacher thought schools would respond positively to beginning teachers with a specialisation in science. She noted how she saw "the younger teachers … sort of coming through [university] and leading it [science education]" (p. 35) while she was on professional experience.

Turning to look at the primary specialisation stimulus paper there is a list of responsibilities that might be expected of this type of educator. Many of these surpass the Graduate career stage of the Australian Professional Standards for Teachers (APSTs). Phrases from the stimulus paper such as *"implement …"* and *"work with colleagues* to …" (AITSL, 2017, pp. 3–4, emphasis added) mirror language from the descriptions of a Proficient or Highly Accomplished teacher where teachers "apply", "model", and "support colleagues". Other phrases lifted from the stimulus document (AITSL, 2017) in relation to science such as *"critique* science curriculum, … pedagogy and practices" (p. 3, emphasis added) and *"evaluate* and *improve* science teaching and learning" (p. 4, emphasis added) are commensurate with Lead Teachers for whom the APSTs use words such as "evaluate", "monitor", and "review" to describe their performance relative to career stage.

To fully answer this question, we need to look beyond primary specialisations in ITE and look at what's happening in schools in Australia and internationally. Although the international literature shows that specialisation is not a common approach (Mills et al., 2020), looking sideways to instructional coaches and specialist teachers we see that these roles are often filled by teachers with lots of previous experience, higher qualifications, and the capacity to be leaders. In the case of instructional coaches whose role is to facilitate co-learning through collaborative and cyclical planning, teaching, and reflection, Campbell and Griffin (2017) inform us that the coach is more knowledgeable and experienced than the classroom teacher. These authors suggest this is very important because the coach may be called upon to support curriculum and resource development, model lessons for teachers, and provide professional learning workshops (Mills et al., 2020). The international literature indicates the most common form of preparation for these educators is through postgraduate courses at university generally focused on content knowledge, pedagogical content knowledge, and leadership knowledge and skills (Mills et al., 2020). In Australia, more often than not, these educators are appointed by their school's principal or are self-appointed (Mills & Bourke, in preparation). The reasons for this can be because they have additional qualifications in science, for example a Bachelor's or Master's degree or because they have a personal interest in or passion for teaching science.

Qualification in science was a driver for only two out of five pre-service teachers in Mills and Bourke's (2020) study. The other three had negative experiences with science. One pre-service teacher described "I chose based mainly off my own experience [at school]" (p. 32). They recalled their science education "was out of a textbook" and there was "not a lot of inquiry" and "not a lot of experimentation" (p. 32).

Another pre-service teacher recalled how "the teacher would write things up on the board" and students "didn't really do much" (p. 32). This sentiment was echoed by yet another pre-service teacher stating they "picked science … in high school … but had a really bad science teacher so … dropped out" (p. 32). They explained further "… when I learnt science it was more transmission … so the teacher just talked, and we wrote notes" (p. 32). These pre-service teachers wanted to do better for future generations of children who had an interest in science.

In response to the question "Who is a teacher with a science specialisation and who can/should have specialised knowledge?", it seems the answer is anyone. Pre-service teachers do not appear to have much guidance when choosing an area of specialisation, some choosing science because they enjoyed or were good at it at school and others choosing science for the exact opposite reason. Given the variability in who are teachers with a science specialisation, the begging question is why are these educators needed in the first instance? This question is answered in Theme 3 next.

4.5.3 Theme 3: Student Enrolment and Engagement Versus Achievement (Why)

This theme considers why teachers with a science specialisation are needed (or not) in primary education. Looking at the political discourse around this topic, the reason given is that primary teaching is too demanding because teachers are responsible for all curriculum learning areas affecting depth of content knowledge and quality of teaching. This is problematic because it positions primary school teachers in a deficit discourse, indicating they do not have the specialist knowledge and skills to teach certain subjects. In addition, children see science being taught by specialist teachers and this passes a message on to them that science is not for everyone and you have to be a special type of person to enjoy or be successful at science. We have already seen evidence of this tension in Theme 1 where a binary between generalist/specialist is set up, with specialists forwarded as more desirable to enhance quality in primary education. To unpack what Carol Bacchi calls the "problem representation" of primary specialisation we find inconsistencies in the logic of arguments presented in policy and sometimes research literature (Bourke et al., 2020). This is related to teacher quality—how it is thought about, spoken about, and brought into existence as reality or truth. We also find that alternative solutions to the problem representation are closed off.

Tracing the origins of primary specialisation in ITE reveals a problem representation related to quality. However, this oscillates between quality defined as student enrolment and engagement in science and quality defined as student achievement in science. Following Bacchi (2009), "concepts [like quality] abstract labels that are relatively open-ended … people fill them with different meanings" (p. 8). Quality in the TEMAG report in 2014 focused on enrolment, engagement, and interest in science, whereas quality in the subsequent AITSL publications focused on student achievement. In both documents it is assumed that student disengagement in the early formative years has led to falling numbers in the senior years, even though writers such as Kennedy et al. (2014) attribute this decline in science to other factors such as subject offerings in senior years or university prerequisites. Likewise, cherry-picked statistics and data are used to defend primary specialisations, for example data from PISA and TIMMS have been used to steer specialisations towards maths, science, and literacy.

Regardless of whether the problem representation is tied to student engagement or achievement, the conceptual logic of this argument is problematic as it assumes a simplistic continuity from teacher quality to impact. Biesta (2010) argues that interventions to increase quality which are premised on if "we do A, B will follow" (p. 494) will not work in education, as there are many factors to consider in the learning process (cf. Kennedy et al., 2014). Therefore, it cannot be assumed that introducing primary specialisations will improve confidence, engagement, student enrolments, or student outcomes. Moreover, Biesta (2015) contends that defining quality in education as impact on achievement/student outcomes alone is one-dimensional, negating the other purposes of education like socialisation (ways of being and doing) and subjectification (student as person). This is true of science education whose purpose for a long time has been scientific literacy, defined by Leonie Rennie as "Knowing science as a way of thinking, finding, organising and using information to make decisions … about the environment and their own health and well-being" (Rennie, 2005, p. 10). This view of school science is potentially challenged by primary specialisations, which are underpinned by the assumption that this is specialised and not an everyday field.

Further tension we see in primary specialisation policy is the closing off of alternative solutions to the problem representation. As pointed out by Bourke et al. (2020), Australia has a crowded curriculum with primary school children expected to learn too much. This is not represented as the problem in political discourse, which focuses on teacher quality instead. Also, if there is in fact a problem with teachers' science knowledge and skills, rather than requiring additional study on top of their ITE, perhaps the volume of learning devoted to science in ITE courses could be increased, as we have just seen in Australia with teaching of early reading (AITSL, 2022).

By way of triumph or silver lining there appears to be minimal harm in primary specialisations, with the pre-service teachers in Mills and Bourke's (2020) study reporting science specialisation modules enhanced their knowledge about pedagogical approaches: "it [the specialisation] gave me a greater understanding … into teaching science to children" (p. 31). One pre-service teacher noted how she developed a repertoire of science teaching strategies completing the specialisation, saying "I can use … different pedagogies to teach science it doesn't have to be through transmission" (p. 31). She noted that her new practices were "supported by evidence" because "the online modules provided research literature showing what works in science teaching" (p. 31). Conversely, primary specialisations do restrict pre-service teachers' options to take subjects in other areas because space in their program is taken up by specialisations.

The next theme deals with teachers with a specialism broadly. Although this section is about how these educators work in schools and classrooms, recommendations for ITE are given.

4.5.4   Theme 4: High Versus Low Status in Schools (How)

This theme answers the question "How do teachers with a specialisation work successfully in schools?" This section is somewhat shorter than the previous sections because pre-service teachers with a specialisation are only just graduating and commencing classroom roles. Here we look sideways to the hybrid practices that are presently occurring in classrooms. While the roles and responsibilities of teachers with a specialisation are well-defined, described at the beginning of this chapter, there is a limited understanding about the conditions that may enable or constrain their work. Drawing on Mills and Bourke's (in preparation) study with primary school teachers who identified as having a science specialism, the perceived value of science/STEM and specialist teachers were important enabling or constraining conditions depending on whether or not they are present. These conditions, what Kemmis and Grootenboer (2008) call arrangements of practices or "practice architectures", were so important that they actually enabled or constrained other practice architectures. These are called "*foundational practice architectures*": they lay the foundations for further enabling practice architectures (Mills & Bourke, in preparation).

The perceived value or status placed upon science/STEM education was identified as an enabling condition by two participants in Mills and Bourke's (in preparation) study. One teacher explained that while her school had placed high importance on mathematics/numeracy, but not STEM, this had changed with the inclusion of STEM in the school's strategic plan. This teacher described how at first "we were just pushed to the side a bit" and "just left to do what we want to do", before the new school plan which included STEM education "shifted focus" creating a climate where "the teachers talk highly of STEM as a subject". The second teacher described how her school's principal "values science as a specialist subject" and appointed her to a school-funded specialist teacher role. These teachers' experiences illustrate a practice landscape comprising discourses wherein STEM education is valued. While this is formalised in a policy document at the first teacher's school, the school's culture is cultivated by the principal at the second teacher's school. Either way, the high status afforded to science/STEM has enabled the teachers' work as a specialist STEM teachers.

The way teachers with a specialism are spoken about and understood appears to be very important. In Mills and Bourke's (in preparation) study, most science/STEM specialist teachers felt like this constrained their work. One teacher explained, "I think we're the extras … there's certainly more focus on classroom teachers …". The specialist teachers in this study had ill-defined job roles and often picked up additional responsibilities like playground duties. Participants had varied job titles such as "Science Specialist Teacher", "STEM Teacher", and "Science Co-teacher".

Although this positions them as specialists or experts, it at the same time denies them working conditions enjoyed by "Specialist Teachers" which in Queensland are restricted to music, physical education, and/or LOTE learning areas. As expressed by one teacher, this often meant missing out on additional specialist preparation and coordination time (i.e., non-contact time): "Yeah so we're called STEM teachers because we're not classed as a specialist … they have more non-contact time and things like that". In summary, a school culture where specialist teachers are devalued constrained their specialist teacher work.

4.6 Conclusions

In summary, this chapter has mapped the what, who, why, and how of primary science specialisation in Australian ITE. Weaved throughout the analysis of five empirical studies about this group of educators are binaries creating both tensions and opportunities for triumphs. In conclusion the authors return to the three questions guiding this book: (1) What does primary science teacher education look like in practice in Australia?; (2) What are the triumphs in the approaches used?; and (3) What are the inherent tensions in these approaches?

What does primary science teacher education look like in practice in Australia? It remains largely unknown how Australian universities are enacting primary special-isations although it appears that many universities offer a specialisation in science (Main et al., 2021). Specialisations appear to be a mix of content knowledge and pedagogical content knowledge units, which in some cases may be taught by science faculties. The volume of learning and its sequencing differs from university to univer-sity. There are very few illustrations of what this looks like in practice, however what is known is that regardless of how primary specialisations are enacted they must adhere to AITSL's criteria of advanced content knowledge and pedagogical content knowledge and highly effective classroom practice.

This chapter has also revealed that primary science teacher education in terms of primary specialisations does not necessarily reflect the practices occurring in schools. The global research literature informs us the most common and effective approach to specialism in primary schools is instructional coaches who by definition are senior experienced teachers capable of leading and affecting change in their more junior colleagues' teaching practice. Also evident in the international literature as well as calls from professional bodies are calls for specialist teachers who are solely responsible for learning and teaching in one subject, for example science. The primary specialisation model is unique in that it targets pre-service teachers (i.e., novices) and demands their teacher education covers all learning areas in addition to specialised knowledge and practice in one of these areas. While there are examples of primary school teachers co-teaching or sharing classes based on their interest in STEM, it is this seemingly unique model of preparing such educators that causes most tensions.

What are the triumphs in the approaches used? To move forward this chapter proposes several recommendations for ITE that could be explored in the future. Given the justifications for primary specialisations in ITE have been de-legitimised (e.g., Bourke et al., 2020), the authors would like to see specialism as a postgraduate award such as a Graduate Certificate or Masters of Education in line with practices already occurring overseas and nationally. In the most successful model these educators would work like instructional coaches, rather than specialist teachers which despite massive financial investments from governments in Australia have had seemingly little impact. The Queensland project of Science Facilitators do not have a publicly available final report and despite commentators saying Victoria's science specialists program is quite successful there appears to be only one peer-reviewed research evaluation study (Herbert et al., 2017). Thinking about actions that can be taken in ITE specifically, if teachers need further expertise in science it makes sense to increase the volume of learning required for science in the ITE Accreditation Guidelines, which for undergraduate programs is currently half of the mandated time for Mathematics/numeracy and half again the mandated time for English/literacy.

What are the inherent tensions in these approaches? The binaries discussed in this chapter revealed several tensions. There remains a lack of information and research about who can and should be a teacher with a science specialisation: Is this something that is decided based on previous experience and qualifications or personal interest in science? Moreover, what exactly is the role of a teacher with a specialisation is yet to unfold. These educators are entering a new profession as a novice while at the same time being a leader or champion within the school demonstrating exemplary knowledge and skills to teach science. This lack of clarity in roles and responsibilities means that these teachers are employed in schools to do various work. Quite often this can align with the work of a specialist teacher, however they are not given this official title which means they miss out on additional non-contact time that is often used for planning and resource preparation and has been linked with greater student achievement in one US study (Brobst et al., 2017). If they are not valued and well supported by their school administration they can also be misused, for example filling vacant time slots in the school's timetable or picking up playground duties.

References

Association for Science Education (ASE). (2019). *Why subject knowledge must be at the heart of teachers' early career framework*. https://www.ase.org.uk/news/why-subject-knowledge-must-be-heart-of-teachers-early-careerframework

Australian Institute for Teaching and School Leadership (AITSL). (2022). *Accreditation of initial teacher education programs in Australia: Standards and procedures*. http://www.aitsl.edu.au/docs/default-source/initial-teacher-education-resources/accreditation-of-ite-programs-in-australia.pdf

Australian Institute for Teaching and School Leadership (AITSL). (2020). *Primary specialisation (program standard 4.4): Guidelines*. https://www.aitsl.edu.au/docs/default-source/default-document-library/aitsl_primary_specialisation_guidelines_2020.pdf?sfvrsn=7f15fe3c_2

Australian Institute for Teaching and School Leadership (AITSL). (2017). *Primary specialisation: Graduate outcomes stimulus paper*. https://www.aitsl.edu.au/tools-resources/resource/primary-specialisation---graduate-outcomes-stimulus-paper

Bacchi, C. (2009). *Analysing policy: What's the problem represented to be?* Pearson Australia.

Biesta, G. J. (2010). Why 'what works' still won't work: From evidence-based education to value-based education. *Studies in Philosophy and Education, 29*(5), 491–503.

Biesta, G. J. (2015). What is education for? On good education, teacher judgement, and educational professionalism. *European Journal of Education, 50*(1), 75–87.

Bourke, T., & Mills, R. (2022). How teacher educators do policy: Enacting primary specialisations. In T. Bourke, D. Henderson, R. Spooner-Lane, & S. White (Eds.), *Reconstructing the work of teacher educators* (pp. 31–50). Springer.

Bourke, T., Mills, R., & Siostrom, E. (2020). Origins of primary specialisation in Australian education: What's the problem represented to be? *Australian Educational Researcher, 47*, 725–740.

Braun, V., & Clarke, V. (2006). Using thematic analysis in psychology. *Qualitative Research in Psychology, 3*(2), 77–101.

Brobst, J., & Markworth, K. (2019). Elementary content specialization: Perspectives on perils and promise. *School Science & Mathematics, 119*(7), 369–381. https://doi.org/10.1111/ssm.12362

Brobst, J., Markworth, K., Tasker, T., & Ohana, C. (2017). Comparing the preparedness, content knowledge, and instructional quality of elementary science specialists and self-contained teachers. *Journal of Research in Science Teaching, 54*(10), 1302–1321. https://doi.org/10.1002/tea.21406

Cambridge Dictionary [online]. (2022a). *Specialization*. https://dictionary.cambridge.org/dictionary/english/specialization

Cambridge Dictionary [online]. (2022b). *Assemblage*. https://dictionary.cambridge.org/dictionary/english/assemblage

Campbell, C., & Chittleborough, G. (2014). The 'new' science specialists: Promoting and improving the teaching of science in primary schools. *Teaching Science, 60*(1), 19–29.

Campbell, P., & Griffin, M. (2017). Reflections on the promise and complexity of mathematics coaching. *The Journal of Mathematical Behavior, 46*, 163–176.

Dailey, D., & Robinson, A. (2016). Elementary teachers: Concerns about implementing a science program. *School Science and Mathematics, 116*(3), 139–147. https://doi.org/10.1111/ssm.12162

Department of Education and Training (DET). (2015). *Action now: Classroom ready teachers—Australian Government response*. Australian Government.

Department of Education and Training (DET) (Victoria). (2019). *STEM in schools*. https://www.education.vic.gov.au/about/programs/learningdev/vicstem/Pages/schools.aspx

Department of Education, Training and Employment (DETE). (2012). *Science spark: Interim evaluation summary*. https://qed.qld.gov.au/det-publications/reports/Documents/evaluation/science-spark-summary.pdf

Harvey, S. (1999). The impact of coaching in South African primary science InSET. *International Journal of Educational Development, 19*(3), 191–205.

Herbert, S., Xu, L., & Kelly, L. (2017). The changing roles of science specialists during a capacity building program for primary school science. *Australian Journal of Teacher Education, 42*(3), 1–21.

Kemmis, S., & Grootenboer, P. (2008). Situating praxis in practice: Practice architectures and the cultural, social and material conditions for practice. In S. Kemmis & T. J. Smith (Eds.), *Enabling praxis: Challenges for education* (pp. 37–62). Sense Publishers.

Kennedy, J., Lyons, T., & Quinn, F. (2014). The continuing decline of science and mathematics enrolments in Australian high schools. *Teaching Science, 60*(2), 34–46.

Levy, A., Jia, Y., Marco-Bujosa, L., Gess-Newsome, J., & Pasquale, M. (2016). Science specialists or classroom teachers: Who should teach elementary science? *Science Educator, 25*(1), 10–21.

Main, S., Byrne, M., Scott, J., Sullivan, K., Paolino, A., Slater, E., & Boron, J. (2021). Primary specialisations in Australia: Graduates' perceptions of outcome and impact. *Australian Educational Researcher*. https://doi.org/10.1007/s13384-021-00496-y

Markworth, K., Brobst, J., Ohana, C., & Parker, R. (2016). Elementary content specialization: Models, affordances, and constraints. *International Journal of STEM Education, 3*, 1–19. https://doi.org/10.1186/s40594-016-0049-9

Markworth, K., Brobst, J., Parker, R., & Ohana, C. (2018). Exploring elementary content specialization: Benefits and cautions, pitfalls and fixes. *NCSM Journal of Mathematics Education Leadership, 19*(2), 3–11.

Mills, R., & Bourke, T. (in preparation). *What practice architectures enable and/or constrain primary school teachers with a science/STEM specialism?*

Mills, R., & Bourke, T. (2020). Primary specialisation in Australian education: Pre-service teachers' lived experiences. In J. Fox, C. Alexander, & T. Aspland (Eds.), *Teacher education in globalised times* (pp. 23–40). Springer.

Mills, R., Bourke, T., & Siostrom, E. (2020). Complexity and contradiction: Disciplinary expert teachers in primary science and mathematics education. *Teaching and Teacher Education, 89*, 1–12. https://doi.org/10.1016/j.tate.2019.103010

National Science Teaching Association (NSTA). (2017). *NSTA position statement: Science teacher preparation*. https://www.nsta.org/about/positions/preparation.aspx

National Science Teaching Association (NSTA). (2018). *NSTA position statement: Elementary science education*. https://www.nsta.org/about/positions/elementary.aspx

Nickerson, S. (2010). Preparing experienced elementary teachers as mathematics specialists. *Investigations in Mathematics Learning, 2*(2), 51–68.

Pezaro, C. (2017, February 13). Specialist science and maths teachers in primary schools are not the solution. *Australian Association for Research in Education*. https://www.aare.edu.au/blog/?tag=specialist-teachers-in-primary-schools

Poland, S., Colburn, A., & Long, D. E. (2017). Teacher perspectives on specialisation in the elementary classroom: Implications for science instruction. *International Journal of Science Education, 39*(13), 1715–1732. https://doi.org/10.1080/09500693.2017.1351646

Rennie, L. (2005). Science awareness and scientific literacy. *Teaching Science, 51*(1), 10–14.

Roach, W., & Wendt, J. L. (2022). An exploration of the use of science specialists and elementary students' science achievement. *Journal of Mathematics and Science: Collaborative Explorations, 18*(1). https://doi.org/10.25891/q2fm-2b67

Schwartz, R., Abd-El-Khalick, F., & Lederman, N. (2000). Achieving the reforms vision: The effectiveness of a specialists-led elementary science program. *School Science & Mathematics, 100*(4), 181–193. https://doi.org/10.1111/j.1949-8594.2000.tb17255.x

Teacher Education Ministerial Advisory Group (TEMAG). (2014). *Action now: Classroom ready teachers*. https://www.aitsl.edu.au/tools-resources/resource/action-now-classroom-ready-teachers

Webel, C., Conner, K., Sheffel, C., Tarr, J., & Austin, C. (2017). Elementary mathematics specialists in "departmentalized" teaching assignments: Affordances and constraints. *The Journal of Mathematical Behavior, 46*, 196–214.

Chapter 5
What Are the Possibilities and Challenges Inherent in STEM for Primary Science Teacher Education?

As Science, Technology, Engineering, and Mathematics (STEM) education becomes more and more of a focus globally, I see an opportunity for science education to evolve. Primary science teacher education can leverage from the opportunities created by STEM education to create even better teachers and therefore more students who are well-equipped for the challenges of the future.
Linda

5.1 Introduction

There is a Science, Technology, Engineering and Mathematics (STEM) 'crisis' which involves a shortage of students—the next generation, who will be responsible for our futures—studying STEM (Office of the Chief Scientist, 2014). One method to address the STEM crisis is to look at the way STEM is taught in schools and to develop educators' skills in teaching STEM-based curriculum to be more engaging and innovative to spark interest in students for future employment within STEM (Department of Education, 2016).

As discussed in the previous chapter, the introduction of STEM specialists along with other National and State government initiatives has influenced the ITE sector. For example, STEM has been embedded into undergraduate and postgraduate ITE courses either as a stand-alone subject or within the science or digital technology units. In this chapter, approaches to primary school STEM education are discussed with a view to critically analyse affordances and constraints for primary science teacher education. Australia's average achievement in reading, mathematics and science has experienced a pattern of long-term decline with recent Programme for International Student Assessment (PISA) results showing the average achievement of an Australian 15-year-old in 2018 was almost one full year of schooling behind in science compared to in 2006 (ACER, 2019). If our students are not achieving in science, then the capacity for future Australian discoveries and innovation may be challenged (Tytler, 2007). Primary teachers need to be equipped with the tools (time, access to experts and resources as well as a mindset) to be able to provide learning experiences conducive to innovative thinking (Forbes et al., 2021). Therefore, primary science teacher education needs to cater for these needs.

As the Australian Chief Scientist at the time, Ian Chubb, noted "international research indicates that 75% of the fastest growing occupations now require STEM skills and knowledge" (Office of the Chief Scientist, 2014, p. 2). Around 85% of the jobs that today's learners will be doing in 2030 haven't been invented yet (Institute for the future, 2017, p. 14). Primary school students in the classroom today will be the workforce in 2030 and beyond. It is quite a challenge to prepare students for careers that may not exist in present time.

The term STEM is used as an acronym for the disciplines of Science, Technology, Engineering, and Mathematics taught and applied either in a traditional and discipline-specific manner or through a interdisciplinary, interconnected, and integrative approach. Both approaches are outcome-focused and aim to solve real-world challenges (Siekmann & Korbel, 2016). An interesting article titled 'STEM, STEM Education, STEM Mania' (Sanders, 2009 cited in Forbes et al., 2021) makes the important point that STEM refers to the fields in which scientists, technologists, engineers, and mathematicians work, while teachers in the fields of STEM are 'STEM educators working in STEM education'. Therefore, when referring to teaching STEM in schools, the term 'STEM education' should be used, not simply 'STEM' (Forbes et al., 2021).

While STEM education and training establishes relationships between the four disciplines with the objective of expanding people's abilities by supporting technical and scientific education with a strong emphasis on critical and creative-thinking skills (Siekmann, 2016), Finkel (2018) attests that there is a need for *real experts, who can lift you above the generic thinking of everyone else.* This creates an opportunity for primary science education in the teaching of the specific discipline of science as well as with integrating technology, mathematics, and engineering design into the primary classroom. Teachers need to be prepared for the future classroom by not only being able to teach the subject-specific knowledge and skills but also apply these in an integrated and inspirational manner to create scientifically and technologically literate citizens who can critically examine/understand/respond to and improve the world around them (Siekman & Korbel, 2016). STEM education offers the chance to make real-world connections by integrating the knowledge and skills of the four disciplines in a meaningful way.

This chapter will start with a brief comparison of the different definitions of STEM skills and how these relate to primary science teacher education. An exploration of both formal and informal STEM education is used to describe some of the opportunities and challenges inherent to primary science teacher education. A case study is used to explore in more detail the ways in which STEM education can be used in pre-service science teacher education, including the use of networks and partnerships as a vehicle to promote primary science. This chapter concludes with some recommendations for what primary science teacher educators can be doing better to prepare future primary teachers for STEM education.

5.2 STEM Skills Definitions

> Understanding that children need a particular set of skills to live constructively in the 21st century is not the same as an in depth understanding of what those skills are, how they develop, or how to teach them. (Care et al., 2017, p. 2)

As Finkel (2018) implied in the opening quote of this chapter, STEM subjects, in particular science, need to be taught distinctly from STEM skills. He has compared the mastering of the discipline to learning to play an instrument in an orchestra. One cannot possibly play successfully in an orchestra without first learning to master their instrument. The term 'STEM skills' is used in a broad range of contexts, for example, by governments, industry bodies, education providers, and public media.

Generating a list of those skills considered to be 'STEM skills' is no easy feat. Descriptions of STEM skills usually include a combination of technical job-specific skills and advanced cognitive skills (Siekmann, 2016). STEM skills can also be confused with employability skills and twenty-first century skills. The terminology used for the different skills varies amongst countries. For example, in Australia we use terms such as key competencies, employability skills, generic skills, and general capabilities (Weldon, 2019). In Canada, the terms employability skills and core competencies are used (Weldon, 2019). The European Parliament uses key competences while in New Zealand these are essential skills. According to the OECD, the terms are key competencies, global competencies, and twenty-first century skills, while UNESCO calls them transversal competencies (Weldon, 2019). These varied terms and lack of agreement globally can create confusion and makes it difficult to agree on what exactly STEM skills are.

In Australia, although there are many government reports outlining statistics around Australia's STEM workforce, STEM careers, and STEM qualifications, the definition of the actual skills referred to as STEM skills is somewhat sporadic. Generally, STEM involves inquiry-based learning, solving real-world problems, creating partnerships, and working in teams. The STEM skills required to achieve these include collaboration, creativity, critical thinking, and communication. These skills are also included in the primary science curriculum. While these skills are important, the future of the nation is arguably in peril if we don't teach our children high level discipline specific knowledge as well. In the primary science classroom, this means far more than building a tower from spaghetti and marshmallows and calling it STEM but rather explaining the scientific concepts involved and using scientific methodology to test different variables to create a result and solve a problem. Teachers need to develop STEM skills in the primary science classroom and pre-service science teacher education needs to ensure they have the knowledge and skills to do so.

Siekmann and Korbel (2016) define STEM skills primarily as technical skills as opposed to higher-order thinking skills. They state that STEM skills are a combination of the ability to produce scientific knowledge, supported by mathematical skills, to design and build (engineer) technological and scientific products or services (Siekmann & Korbel, 2016). It is interesting to note that in most definitions of STEM as an integrated approach, science is the backbone of the disciplines while mathematical *links* can be made, technology *tools* can be utilised, and engineering design *processes* can be employed. Therefore, STEM education is important within science, providing the central core from which to build.

The term 'twenty-first century skills' usually refer to the skills desired currently by employers, although they are often confusingly assumed to be STEM skills (Siekmann, 2016). While there is broad agreement that students need skills different from those perhaps taught to previous generations, and that foundational skills such as writing, critical thinking, self-initiative, group collaboration, and technological literacy are essential to success in higher education, modern workplaces and adult life, there is still a great deal of debate about twenty-first century skills—from identifying the most important skills and how these skills should be taught, to their appropriate role in public education (Great Schools Partnerships, 2014). From the U.S. National Science board (2015), these include complex problem solving, technology design, and programming as well as deductive and inductive reasoning, mathematical reasoning, and facility with numbers. Instead of specific subject knowledge, twenty-first century skills are ways of thinking, ways of working, and ways of living (Australian Curriculum, Assessment and Reporting Authority [ACARA], 2022). The Australian Curriculum includes seven general capabilities:

- Literacy,
- Numeracy,
- Information and communication technology (ICT) capability,
- Critical and creative thinking,
- Personal and social capability,
- Ethical understanding, and
- Intercultural understanding (ACARA, 2022).

Since twenty-first century skills involve ways of thinking and ways or working, teachers need to be able to create lessons that promote critical thinking, communication, collaboration, and creativity. Figure 5.1 demonstrates the comparison of some of the STEM skills defined by the Australian Government Office of the Chief Scientist, common twenty-first century skills and skills as defined by the Australian Curriculum: Science specifically The Australian Curriculum general capabilities include critical thinking, problem solving, communication and collaboration. This figure demonstrates that skills such as communication align directly with the Australian Curriculum science inquiry skills already evident in the primary science classroom and central to preparing primary science educators.

In 2015 there was an agreement made by all Australian state and territory Education Ministers to implement a National STEM School Education Strategy (Education Council, 2015) for the period spanning 2016–2026. Two high-level goals for the strategy were set:

(1) That all students finish school with strong foundational knowledge in STEM and related skills; and
(2) That students are inspired to take on more challenging STEM subjects.

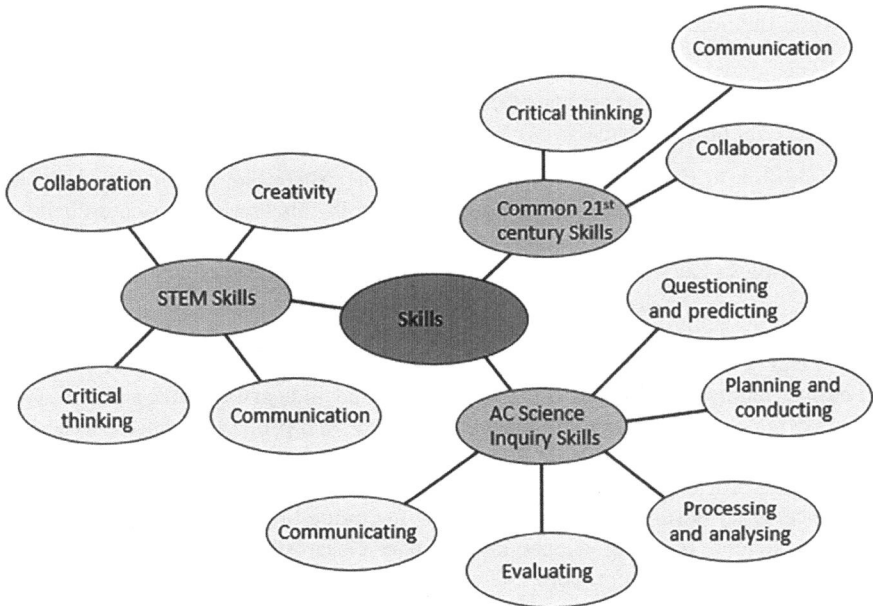

Fig. 5.1 Comparison of common STEM skills, twenty-first century skills and ACS skills

For primary science education, this means pre-service and in-service teachers will play key roles in achieving these goals, interpreting the various curriculum and syllabus documents, and then designing and implementing innovative, relevant, and engaging learning opportunities for students (Forbes et al., 2021). Primary science teacher educators need to prepare pre-service teachers for these roles by embedding STEM knowledge, skills, and pedagogies into initial teacher education courses, specifically in primary science.

5.3   Formal STEM Education

The Australian Curriculum is made up of eight key learning areas. These are:

- English,
- Mathematics,
- Science,
- Health and Physical Education,
- Humanities and Social Sciences,
- The Arts,
- Technologies, and
- Languages.

Within the Science key learning area there are three strands:

- Science understanding (SU),
- Science as a human endeavour (SHE), and
- Science inquiry skills (SIS).

In addition to the eight key learning areas, the Australian Curriculum is made up of seven general capabilities (as listed earlier in this chapter) to be embedded across the eight key learning areas. In addition, the Australian Government offers nine principles (see Table 5.1) that should be applied in STEM education (ACARA, 2022).

The nine principles identified above offer some ways in which innovation and STEM can be integrated into the primary science classroom. Using problem-based scenarios, teamwork, student-centred learning, and industry experts are already a common feature of learning and teaching in the formal primary science classroom. With this in mind, approaches to primary science teacher education should include opportunities for engaging with student-centred learning, teamwork and problem-based scenarios during tutorials as a means for modelling these practices for pre-service teachers. Teaching science in a primary classroom from a STEM perspective draws upon these approaches. Current best practice when teaching science is through the use of student-centred approaches and collaborative groups (Woolcott & Whannell, 2017).

Forbes et al. (2021) combined processes found in mathematical modelling, scientific inquiry, technology design and engineering design to develop five steps for embedding STEM into the formal curriculum through a project-based learning (PBL) model. These steps are:

1. Project identification, challenge or issue;
2. Curriculum connections;
3. Assessment ideas;
4. Teaching and learning processes; and
5. Reflection.

Within the teaching and learning step, the following five phases have been developed:

- Phase 1: *ASK*—defining and understanding the project requirements.
- Phase 2: *IMAGINE*—investigating/researching about the project.
- Phase 3: *PLAN*—designing/planning the preferred solution.
- Phase 4: *CREATE*—producing the preferred solution.
- Phase 5: *IMPROVE*—delivering and evaluating the solution.

The phases within learning and teaching during an integrated STEM PBL model as outlined above provide one way to approach integrated STEM education projects in the primary curriculum. This process has similarities to the science inquiry skills regularly used in the primary science classroom. The *Garden Challenge* project is an example of how to implement STEM education in a primary school classroom

Table 5.1 Principles in STEM education (ACARA, 2022)

Principle	What is it?	Why is this important?	Examples
Use inquiry-based learning	Inquiry-based learning is an education approach that focuses on investigation and problem-solving	Students learn key STEM and life skills through inquiry-based learning: social interaction, exploration, argumentation, comfort with failure	Build active learning into teaching practices through problem-based scenarios to encourage students to think critically
Solve real-world problems	Students tackle real-world STEM problems from industry and the community	Demonstrates relevance of STEM; can enhance student motivation and interest and highlight career opportunities	Partner with a local business and get students to work on a project that solves a real problem to see what they come up with
Teach integrated STEM learning	Integrated STEM learning combines the subject matter of two or more STEM subjects into a joint learning experience	Supports cross-disciplinary STEM skills; can enhance student interest	You can teach science using an engineering process (design-based learning)
Equip and empower teachers	Equipping and empowering teachers means providing them with the right resources (e.g. high-quality professional learning opportunities, up-to-date technology) and skills to teach best practice STEM education	Teachers have the greatest influence on in-school achievement and engagement in STEM education	Connect a STEM teacher with a STEM mentor from a local business
Create partnerships between schools, industry and community	Schools, businesses and other organisations create STEM education initiatives to improve student outcomes	Exposes students to the workplace, inspires enthusiasm about STEM and enhances and complements curriculum	Choose some partners to work with on a STEM problem. Reach out to schools, industry, museums, local councils and government
Engage parents and families	Encourage parents and guardians to be active in their children's education	Improves aspiration, enrolment, achievement and belief in importance of STEM education	Invite parents and families to a STEM exhibition day to show them all the exciting things students are working on

(continued)

Table 5.1 (continued)

Principle	What is it?	Why is this important?	Examples
Use technology as an enabler	Selective use of technology to support high-quality teaching and learning	Accelerates student learning, increases confidence and ability in using technology	Get students to program a technology instead of showing them what something does
Differentiate for different levels	Learning is tailored to the needs and abilities of individual students	Supports all students' needs, regardless of starting point	Assess student capability formally and informally so lessons can be tailored
Link STEM education to twenty-first century learning	STEM education is intentionally linked to the development of twenty-first century skills such as critical thinking, creativity and collaboration	twenty-first century skills are highly valuable for students' future careers	Encourage teamwork and healthy debate. Let students 'play' with their STEM subject matter

through a 'science concept lens' (Forbes et al., 2021). The project provides opportunities for students to use age-appropriate knowledge of Science Understanding (SU), Science Inquiry Skills (SIS), and Science as a Human Endeavour (SHE) in an authentic way by designing and building a garden in the school. This allows students to develop knowledge and understanding, as well as skills and values, rather than simply engage in 'one-off activities' (Forbes et al., 2021). Projects such as the *Garden Challenge* can be embedded into primary science courses during ITE to provide pre-service teachers with first-hand experiences at implementing STEM projects using a science context.

One of the challenges faced in primary science teacher education is teacher capability since most teachers in the primary sector are educated as generalist teachers and may not have the depth of knowledge in STEM-specific disciplines. Although teachers are educated in certain areas, they cannot bring the same depth of understanding of practicing or trained STEM education professionals (Tytler et al., 2018). Primary science teacher educators face the challenge of preparing pre-service teachers with content-specific knowledge and skills particularly when students arrive with a diverse range of backgrounds and experiences, ranging from a very low interest and sometimes low self-efficacy in science right through to people with a science degree who are re-skilling to become primary teachers. Primary science teacher educators need to provide opportunities for pre-service teachers to undertake PBL tasks that incorporate the science curriculum as well as mathematics and technologies so that practical applications can be applied in their classrooms when they graduate. With a current teacher shortage in Australia, one possible way to include STEM within ITE programming for primary science teachers is by providing industry placements for pre-service teachers to obtain real-world experiences and access to experts in the disciplines. The next section presents some of the benefits and challenges of informal STEM education and implications for pre-service teacher education.

5.4 Informal STEM Education

STEM education is vital to all aspects of the nation's growth including Australia's competitiveness, health and wellbeing, and prosperity. Over twenty years of reports and articles from government, business, think tanks and the media have drawn attention to the STEM learning issue of decreasing numbers of students selecting STEM subjects in secondary school and at university (Timms et al., 2018). Investment into STEM education and research has been recognised as critical to the future success and growth of the country. Providing real-world and contextual applications is fundamental when providing STEM experiences (Nadelson et al., 2013). This requires teachers to draw upon their knowledge, skills, and creativity (Pfeiffer, et. al., in press). STEM experiences in the school context can include informal or outside classroom activities such as after school STEM clubs (Davis et al., 2021). However, limited confidence in approaching a student-centred and inquiry style of learning and teaching is often seen in teachers when teaching STEM curriculum either informally or in the classroom context.

For the purposes of this chapter, informal STEM education refers to STEM experiences that are not necessarily aligned to the Australian Curriculum and are not included in the regular classroom. That is, activities that are not available to every student, only to those who elect to be involved as an additional or extra-curricular activity. Informal STEM education can provide opportunities for students and teachers to engage in science in an applied way. This section discusses the lack of training for teachers in informal settings, some resources available to in-service and pre-service teachers, and some of the benefits and challenges of informal STEM experiences and implications for pre-service teacher education.

Existing studies on teacher learning and progress in informal settings focus on pre-service teachers who have undergone formal teacher education and how being in informal settings can complement their skills for a formal educational setting. Whereas many teachers in informal settings are not required to have had formal educational training. That is, for a teacher to be coordinating an extra-curricular STEM club at a school they may not have any formal STEM education training or professional development. Koch and Gorges (2016, as cited in Kim & Keyhani, 2019) studied several women STEM facilitators working in an informal setting who came from different educational backgrounds and interests. They found that all of the women STEM facilitators experienced a level of professional growth because of their STEM experience. Therefore, allowing pre-service teachers the opportunities to be involved in informal STEM activities during ITE programs could improve their professional growth. This could be achieved by including informal STEM experiences in the educational training of pre-service teachers during their university studies. Research shows that informal STEM settings can encourage teachers to try new teaching methods, and enhance their classroom-relevant competencies such as creativity, social skills, and leadership (Terrazas-Marín, 2018 as cited in Kim & Keyhani, 2019).

There are many STEM resources and informal opportunities available to schools in Australia. The Australian Government has set up an online portal called the STARportal for parents, students and teachers that lists opportunities by areas of interest (Australian Government, 2022). Scoping of the activities listed on the STARportal reveal that most of these are in the sciences, which again demonstrates the advantages and affordances of including STEM education in primary science teacher education, as opposed to other subject areas. Providing pre-service teachers with information regarding sites such as the Australian government STARportal can assist primary science teachers to access ideas for informal STEM experiences. Many organisations including Commonwealth Scientific and Industrial Research Organisation (CSIRO), Queensland Museum, Questacon, Universities, Australian Maths Trust, Education Queensland (STEM Girl Power, Queensland Virtual STEM Academy) offer informal opportunities mainly funded by local, State or Federal Government or industry partners. For primary science teacher educators, there are many opportunities to access funding and resources to embed these informal STEM programs in university courses.

Integrated STEM programs and experiences can increase student engagement in the classroom (Struyf et al., 2019). Pre-service primary teachers need to be exposed to these programs first-hand during their ITE program. Utilising informal STEM experiences such as citizen science programs and accessing resources, such as apps and experts in their fields, allows for school students to apply science content from an informal experience into the classroom. For example, the Port Curtis Harbour Watch program in Gladstone (Department of Education, 2022) allows for citizen science water testing opportunities to be used to learn about habitats and ecosystems. Another example might be to have an excursion to an outdoor education centre as part of the tutorial session. These experiences can be utilised in the university classroom with pre-service teachers as part of primary science teacher education allowing pre-service teachers a real-world lived experience.

Primary science teachers can utilise informal STEM programs including robotics and technology-based programs to motivate students. Using technology as a tool, primary teachers can excite students. Generating local programs and competitions can also provide a conduit for application of curriculum knowledge. Primary science teacher education can embed local programs and competitions in coursework either during tutorials, volunteer hours or as assessment tasks within courses.

> The study of science, technology, engineering and mathematics (STEM) is undoubtedly a gateway to the future for Queensland students. That's why the Palaszczuk Government will invest more than $56 million in STEM-specific initiatives in 2018.
> The Hon. Grace Grace MP, 2018.

Some of the challenges that are faced with informal STEM programs are that the funding of programs is usually short-term and if the teacher leading the program leaves the school, the program tends to cease. Primary science teacher education could include informal STEM programs in courses to allow pre-service teachers to have the confidence to lead programs in their schools. Teachers are often not provided with the skills and knowledge to confidently facilitate informal STEM programs and this gap could be filled through ITE.

By way of example, CQUniversity (a regional university based in the Australian state of Queensland) identified the need to assist regional teachers to engage with STEM and learn new ways of teaching STEM. As a result, the teaching team developed a fourth year (which is the final year of the program) subject in the Bachelor of Education (Primary) called *STEM Education and Engagement*. In addition to this course, STEM has been embedded in the science subject in the Master of Teaching (Primary) to help prepare pre-service teachers to the realities of STEM education in a primary school. For example, one of the assessment tasks involves designing a STEM activity which uses an informal context to provide learning opportunities for all students. This opportunity provides pre-service teachers with an experience in planning, designing, and implementing an informal STEM activity.

The next section presents a case study to explore in more detail the ways in which STEM education can be used in pre-service science teacher education, including the use of networks and partnerships as a vehicle to promote primary science.

5.5 Case Study

ITE programs need to provide pre-service primary teachers with opportunities as well as the skills needed to develop and sustain relationships. This focus is critical as there are many partnership opportunities that can be leveraged for the primary science educator. These opportunities include the Australian Government's Commonwealth Scientific and Industrial Research Organisation (CSIRO), which offers a program called *STEM Professionals in Schools* and can be used to find science experts to connect with pre-service teachers. Professional organisations such as the *Australian Science Teachers Association* (ASTA) can provide primary science teacher educators with resources including access to events (for networking), professional learning (for both primary science teacher educators and pre-service teachers), awards, competitions, webinars, and group projects. These resources can be used to inform primary science teacher educators course development, tutorial content, and assessments. Science museums such as *Questacon* can be also used to provide primary science teacher educators with ways in which to engage pre-service teachers in course materials and keep up-to-date on innovations in science and technology. In Australia, there are *STEM Hub networks* in every State and Territory funded by *Inspiring Australia*, an off-shoot of the Australian Government Department of Industry, Science and Resources. These networks consist of partnerships between industry, community groups, and schools. Although they are named *STEM Hubs*, one of their key tasks is to run events during National Science Week. These networks provide opportunities for primary science teacher educators to form partnerships that can assist with planning tutorials, assessments, and practical experiences for pre-service teachers as well as links to potential partners and resources.

For this chapter, the role that STEM can play in providing real-world context for science learning is explored using a case study from an Indigenous program implemented in Central Queensland called *Buraligim Weiber* (Place of Learning). The program has been developed by the CQUniversity STEM Central team and funded by Australia Pacific LNG. The program involves Year 3 and 4 Aboriginal and Torres Strait Islander students (8–10-year-olds) from local primary schools participating in various activities in a bid to improve engagement, literacy, and school attendance. The school students receive hands-on experiences and excursions, including boat tours of Gladstone Harbour, Quoin Island Turtle Rehabilitation Centre, visits to the Botanic Gardens, Spinnaker Park, and CQUniverity's Coastal Marine Ecosystems Research Centre (CMERC). The program was developed by a team of educators, academics and local industry and community members including Traditional Owners. The program allows the children to learn about the world around them, including impacts on the environment, food and sustainability, local flora and fauna, waterways all through the viewpoint of their connection to Country.

The *Buraligim Weiber* program is an intensive program which involves the Indigenous students attending the university for four hours (part of one day) per week for 20 weeks. While these children have the intensive program, the program has also been implemented in-school for two classes of students, substituting some of the excursions with incursions with community and industry partners. Allowing local and contextual learning through this integrated approach, which includes literacy, numeracy, digital technologies, sciences, and history and social sciences (HASS) learning areas, provides opportunities for the children to apply their learning to real-world experiences. This program is mapped to the Australian Curriculum and is considered a formal STEM program for that reason. The intensive program, however, for the Indigenous students attending the program outside of the school grounds while also being mapped to the formal curriculum is implemented 'off site' and outside the classroom. This gives the program a unique 'informal' flavour. The off-site program provides pre-service primary teachers with the opportunity to volunteer each week through assisting with the implementation of learning and teaching activities.

The *Buraligim Weiber* program includes many outdoor learning experiences. Real-world contexts allow for excursions to industry and local community areas of significance. The off-site program provides the pre-service teachers with first-hand insights into how connections can be made between science concepts and authentic contexts. The program gives the pre-service teacher volunteers opportunities to form relationships by making connections with families, local Indigenous groups, industry, and community experts. These connections provide the pre-service teachers with access to people who can assist with making links between real-world contexts and science curriculum in the primary classroom. It has provided the children, the pre-service teachers, and the primary science teacher educators opportunities to meet real people working in STEM fields and make connections with them on both personal and professional levels. This is a quote from one of the volunteers about the connections between the children and the STEM experts:

> I think it was actually the connections of people that came in, the experts that came in. So the kids… they've connected with Dr [Adam] or Dr [Melissa] and then they've come more than once and they've really connected as well. And that even though they're these scientists, they really connected with them like they're just everyday people and yeah, so I think that was a big thing. [P2]

There are a lot of challenges to overcome in primary education when embedding real-world contexts into the classroom. While STEM education offers the opportunity for primary science teachers to provide real-world contexts to classroom learning, the implementation of the *Buraligim Weiber* program did involve navigating some hurdles. While this case study is one way to offer first-hand experiences to pre-service teachers and improved knowledge and understanding for primary science teacher educators, not every university has the funding to sustain such a large project.

Embedding anything into the classroom involves links to the formal curriculum as discussed earlier. For the *Buraligim Weiber* program, it was critical to have links to the curriculum. This quote is from a teacher:

> Yeah, it's critical. I wouldn't have done it if it wasn't, if we couldn't teach, assess, and report this as a learning area because it's – the curriculum, school, the expectation on our teachers is far too great to take a day a week to play, and that is the reality of it. This needed to have fidelity, this program, and that means that it's not just some fun little thing that kids go off and do and has no connection to school and to their learning. It was critical for me that this is built into part of our curricular, it's part of our whole, we call it level one planning. [P1]

Another way that ITE could utilise STEM opportunities is to provide primary science teachers access to STEM experts in their disciplines. This allows for exposure to real-world contexts to draw from when planning and implementing learning experiences in the classroom. During the *Buraligim Weiber* program, pre-service primary teachers were included in the planning and development of activities in collaboration with science experts as well as being immersed in the excursions and field work. For example, developing rotational activities for the school students with the marine ecologist around marine debris and specifically microplastics.

Networks and partnerships can be a vehicle to providing these contexts. Developing links within the community that includes university and industry-based STEM facilitators enhances the opportunity for engagement both of students and teachers in the STEM education space (Pfeiffer et al., in press). To provide rich STEM learning experiences for students, STEM connections need to be made between partners and students' learning. No one has first-hand experience working in every occupation or field. Teachers need to be able to connect student learning with real-world contexts, despite not necessarily having ever experienced these themselves. This is where partnerships play a vital role in providing teachers and primary science teacher educators with the networks and conversations to link learning environments with real-world contexts (Forbes et al., 2021).

The Review of STEM Education in Queensland State Schools Final report (Queensland Government, 2018) provided six key strategies. One of these was the inclusion of promoting and establishing partnerships between schools, universities, community, and industry. Outlined as the reasoning behind these partnerships was the emphasis on real-world examples for students learning in addition to increasing the capability of teachers to feel "confident about their STEM knowledge and use of effective STEM pedagogical practices" (Queensland Government, 2018, p. 13).

According to the Roadmap to STEM Education in Queensland, a STEM student in 2020:

is connected: to their local community, to the worldwide STEM community, to a central e-STEM learning space, through a variety of technologies (all endorsed for full use in the classroom),

is partnered with industry and universities (access is easy and embedded in the system),

is linked to real world issues,

is engaged in inquiry pedagogy to build their knowledge and skills,

is motivated through autonomy, mastery through progression, contribution,

is living in a world that is increasingly STEM aware,

is collecting data from a variety of sources (local, state, national, global),

is developing choice in their learning activities and learning environments,

is building skills to critically analyse and interrogate data and experiences. (Office of the Queensland Chief Scientist, 2013, p. 5)

Pfeiffer and Tabone (2020) suggest that three critical aspects are needed to develop effective partnerships to enhance STEM education: building capacity, shared visions, and sustainability. These three facets of partnerships can assist with making informed decisions on the STEM initiatives available and which ones involve appropriate expertise, involve people who are reliable and deliver good quality outcomes. Creating partnerships with the various groups available—government, educational organisations, and local industry and community groups—is important in STEM education, as it allows teachers to explore real-world authentic contexts for project ideas (Forbes et al., 2021). Primary science teacher educators can use these aspects to build their own partnerships so that opportunities can be embedded in pre-service teacher education.

Partnerships can have a flow-on effect by improving the knowledge of those involved. That is, primary science teacher educators can increase their knowledge through access to expertise in places such as industry and community organisations. For example, this quote from a teacher educator participant in the *Buraligim Weiber* program shows how being able to attend the excursion and work with industry experts has impacted their knowledge:

Yeah, [my role as a learner] I was picking up new facts and I had – when we went on the harbour cruise at the start, I hadn't seen the [LNG] plants and I learnt a lot of new things about that. And then I can pass that on to the students, not just in the program but in school as well. And yeah, there's just many of the people that came to the program have lived these fascinating careers and lives and yeah, there's so much knowledge to share amongst everybody. [P2]

Partnerships with Traditional Owners are to be encouraged for teachers. Primary science teacher educators need to assist pre-service teachers to gain confidence in building these relationships. This quote from a teacher educator participant in the *Buraligim Weiber* program shows that while it can be challenging to work outside your comfort zone, with the diversity of a team and appropriate support stakeholders with the same goals can work together to produce positive outcomes.

I think as a non-Indigenous, as a white woman, I had to overcome that feeling of what do I know? How can I be promoting this as a great program for Indigenous kids? I'm not Indigenous. I don't have lived experience. For me to get past that feeling, push through that and think we're just going to do it anyway. And I'm doing it with an open and true heart with only good intentions, and if I need to adjust that, that's okay. [P1]

Preparing pre-service primary teachers to promote STEM education and to be leaders in their schools requires primary science teacher educators to remain up to date with STEM careers, STEM skills, informal and real-world learning experiences. ITE programming needs to involve opportunities for partnerships and industry relevant experiences to maximise the benefits, which include improved knowledge and understanding.

5.6 Conclusion

All knowledge is useful. In this report, the word 'science' embraces all ways of knowing. Economics, Indigenous knowledge, technology, mathematics, the social sciences, engineering and the arts are all knowledge ventures and form part of the broader definition of 'science' embodied in this document. (Office of the Queensland Chief Scientist, 2022)

This chapter provided a summary of some of the opportunities that STEM education can bring to primary science teacher education including the use of informal STEM in schools, integrating STEM into the formal curriculum, using real-world contexts for science learning and the ways in which networks and partnerships can be leveraged.

To provide rich learning experiences for students, teachers need to be able to make connections. No one has first-hand experience working in every occupation or field. Teachers need to be able to connect student learning with real-world contexts, despite not necessarily having ever experienced these themselves. This is a challenge but while the jobs of the future may be somewhat unknown, STEM involves solving problems within a focused context. Although the term STEM may go out of fashion, these fundamental skills such as collaboration, problem solving, and interdisciplinary approaches will always remain. This provides possibilities in STEM for primary science education.

Another important point that has not been discussed in this chapter is that ITE relies on in-service teachers acting as mentors during training and in the first years of teaching. Primary science teacher educators therefore also need to be assisting these mentor teachers, particularly with the ever-changing science and STEM landscape, so that effective role models can be used during practical placements for pre-service teachers.

In summary, below are some recommendations for what primary science teacher educators can be doing better to prepare future primary teachers for STEM education.

Recommendations for primary science teacher educators include:

1. Develop and implement STEM courses into the primary education degrees.
2. Embed STEM units as part of the science courses in the primary education degrees.

3. Meet with experts in their fields to inform course development within the primary education degrees.
4. Provide opportunities for pre-service teachers to meet with experts in their fields either as an assessment task within the unit or through networking events.
5. Meet with STEM teachers and schools excelling in STEM education to inform course development.
6. Provide opportunities for pre-service teachers to engage with in-service STEM teachers and other primary teachers and schools excelling in the STEM education space.
7. Participate in conferences and conduct research to inform course development.
8. Provide opportunities for pre-service teachers to join professional organisations by providing details of memberships.
9. Create projects with real-world contexts and links to the Australian Curriculum for pre-service teachers to use at a school, or in tutorials, or as an assessment task.
10. Provide details to pre-service teachers of local and online informal extra-curricular STEM experiences such as STEM clubs and camps for primary school students.
11. Provide practical placement experiences including in STEM workplaces.
12. Support mentor teachers who have STEM specialisations by providing professional development opportunities for collaboration to inform course development.

An expanded vision for what science teaching involves, including STEM, is needed for the future direction of primary science education. The future of primary science education is explored in the next chapter. STEM: is it just good science teaching?

References

Australian Council for Educational Research (ACER). (2019). *PISA 2018: Australian students' performance.* https://www.acer.org/au/discover/article/pisa-2018-australian-students-performance

Australian Curriculum, Assessment and Reporting Authority [ACARA]. (2022). *Links to 21st century learning.* https://www.education.gov.au/australian-curriculum/national-stem-education-resources-toolkit/i-want-know-about-stem-education/what-works-best-when-teaching-stem/links-21st-century-learning

Australian Government. (2022). *STARportal.* https://starportal.edu.au/

Care, E., Kim, H., Anderson, K., & Gustafsson-Wright, E. (2017). *Skills for a changing world: National perspectives and the global movement.* Center for Universal Education at Brookings. https://www.brookings.edu/wp-content/uploads/2017/03/global-20170324-skills-for-a-changing-world.pdf

Davis, K., Fitzgerald, A., Power, M., Leach, T., Martin, N., Piper, S., Singh, R., & Dunlop, S. (2021). Understanding quality learning and teaching in STEM clubs: What does the evidence base tell us? *Studies in Science Education, 59*(1), 1–23. https://doi.org/10.1080/03057267.2021.1969168

Department of Education. (2022). Port Curtis Harbour Watch. https://harbourwatch.eq.edu.au/

Department of Education. (2016). *Advancing education: An action plan for education in Queensland.* https://cabinet.qld.gov.au/documents/2015/Oct/AdvEd/Attachments/AdvEdBooklet.PDF

Education Council. (2015). *National STEM school strategy 2016–2029: A comprehensive plan for science, technology, engineering and mathematics education in Australia.* Department of Education, Skills and Employment.

Finkel, A. (2018). The light in the cave. *The Conversation.* https://theconversation.com/finkel-students-focus-on-your-discipline-then-youll-see-your-options-expand-107440

Forbes, A., Sheffield, R., Pfeiffer, L., & Chandra, V. (2021). *STEM Education in the primary school: A teacher's toolkit.* Cambridge United Press.

Great Schools Partnerships. (2014). *The glossary of education reform: 21st century skills.* https://www.edglossary.org/21st-century-skills/

Institute for the Future. (2017). *The next era of human machine partnerships: Emerging technologies' impact on society & work in 2030.*

Kim, M. S., & Keyhani, N. (2019). Understanding STEM teacher learning in an informal setting: A case study of a novice STEM teacher. *RPTEL, 14,* 9. https://doi.org/10.1186/s41039-019-0103-6

Nadelson, L. S., Callahan, J., Pyke, P., Hay, A., Dance, M., & Pfiester, J. (2013). Teacher STEM perception and preparation: Inquiry-based STEM professional development for elementary teachers. *The Journal of Educational Research, 106*(2), 157–168. https://doi.org/10.1080/00220671.2012.667014

National Science Board. (2015). *Revisiting the STEM workforce: A companion to science and engineering indicators 2014.* National Science Foundation. http://www.nsf.gov/nsb/publications/2015/nsb201510.pdf

Office of the Chief Scientist. (2014). *Science, technology, engineering and mathematics in the national interest: A strategic approach.* Australian Government.

Office of the Queensland Chief Scientist. (2022). *The state of science in Queensland 2022.* Queensland Government.

Office of the Queensland Chief Scientist. (2013). *Roadmap to STEM education in Queensland a strategic roadmap to support primary and secondary science, technology, engineering and mathematics (STEM) education in Queensland.* Queensland Government.

Pfeiffer, L., Bradbury, O., Tabone, K., & Rashleigh, M. (in press). Stimulating Australian STEM education in regional Queensland through a novel university-school-industry partnership. In O. Bradbury & D. Acquaro (Eds.), *International perspectives on school-university partnerships—Research, policy and Practice.* Springer.

Pfeiffer, L. G., & Tabone, K. L. (2020). A case study of a university-industry STEM partnership in regional Queensland. In A. Fitzgerald, C. Haeusler, & L. G. Pfeiffer (Eds.), *STEM education in primary classrooms: Unravelling contemporary approaches in Australia and New Zealand* (pp. 61–78). Routledge. Retrieved from https://www.taylorfrancis.com/

Queensland Government. (2018). *The review of STEM education in Queensland state schools final report.* https://education.qld.gov.au/curriculums/Documents/review-of-stem-education-queensland-state-schools.PDF

Rashleigh, M., Bradbury, O., & Pfeiffer, L. (in press). Connecting informal STEM experiences to real world practices and local applications using an integrated approach across the Australian Curriculum. *Canadian Journal of Science, Mathematics, and Technology Education.*

Siekmann, G. & Korbel, P. (2016). *Defining 'STEM' skills: Review and synthesis of the literature.* NCVER.

Siekmann, G. (2016). *What is STEM? The need for unpacking its definitions and applications.* NCVER.

Struyf, A., De Loof, H., Boeve-de, P. J., & Van Petegem, P. (2019). Students' engagement in different STEM learning environments: Integrated STEM education as promising practice? *International Journal of Science Education, 41*(10), 1387–1407. https://doi.org/10.1080/09500693.2019.160798

Timms, M., Moyle, K., Weldon, P., & Mitchell, P. (2018). *Challenges in STEM learning in Australian schools*. Australian Council for Educational Research.

Tytler, R., Symington, D., Williams, G., & White, P. (2018). Enlivening STEM education through school-community partnerships. In R. Jorgensen, K. Larkin (Eds.), *STEM education in the junior secondary*. Springer.

Tytler, R. (2007). *Re-imagining science education: Engaging students in science for Australia's future*. Australian Council for Educational Research (ACER). https://research.acer.edu.au/cgi/viewcontent.cgi?article=1002&context=aer

Weldon, P. (2019). *Changing priorities? The role of general capabilities in the curriculum*. Camberwell, Australia: Australian Council for Educational Research. https://research.acer.edu.au/nsw curriculumreview/3

Woolcott, G., & Whannell, R. (2017). *Teaching Secondary Science*.

Chapter 6
Where Is Primary Science Teacher Education Headed?

It is, of course, very difficult to predict let alone know the unknowns we face. This reality, understandably, makes it challenging as a teacher educator to prepare future teachers to be classroom-ready. The best we can hope for is to support future primary teachers of science to be curious about the world and open to learning.
Ange

6.1 Introduction: Crystal Ball Gazing

As detailed in the opening chapter and further explored through the subsequent chapters, primary science teacher education in Australia is positioned within a particularly interesting moment in time. There are two key considerations front of mind. First, as the higher education sector continues to dust itself off after the immense challenges it faced during the global pandemic (TEQSA, 2021), many institutions are using this moment to pause and reimagine what the sector might look like going forward (Marmolejo & Groccia, 2022; PwC, 2020). COVID-19 caused higher education institutions to operate differently and now it is proving to be a catalyst for a re-visioning process (Khamis et al., 2021). Second, the global pandemic raised, in the general social consciousness, the importance of scientific literacy and the ability to use evidence-based science knowledge as a lens for informed decision making (Tasquier et al., 2022). This shared experience has foregrounded the important role of science educators in preparing children and young people for an uncertain future, where a solid grounding in science will be vital for making sense of the world and more informed decisions (Durraiappah et al., 2022). With these two considerations in mind, primary science teacher education stands at a fork in the road. Does it stay on the path that maintains the status quo steeped in tradition and certainty or veer into an opportunity to deeply rethink the role and purpose of science education into the future? Given the inherently rhetorical nature of this question, the reality is that this metaphorical horse has bolted and the time for change is now.

© The Author(s), under exclusive license to Springer Nature Singapore Pte Ltd. 2024   93
A. Fitzgerald et al., *Contemporary Australian Primary Science Teacher Education*,
SpringerBriefs in Education, https://doi.org/10.1007/978-981-97-5660-5_6

The Wellcome Trust, a UK-based philanthropic foundation, embarked on a similar thought experiment as detailed above to consider the next steps for primary science education (see: Stubberfield & Barton, 2021). Wellcome's ongoing work in this area has been driven by their guiding premise that "high quality primary science teaching ensures that young people will be better prepared for their futures" (Stubberfield & Barton, 2021, p.4). Their extensive consultation process with a wide variety of stakeholders, including teachers and school leaders, started a number of years prior to the pandemic, but was just starting to surface an emergent theory of change to capture their learning as COVID-19 led to large-scale school closures across both the UK and the world (UNESCO, 2022). Rather than change direction in terms of the overall research vision, a report was developed to capture the learning from this unique moment in time about the overall impact of COVID-19 on schools and science learning and teaching (see: Leonardi et al., 2021). Of particular interest to this chapter, however, are the following three principles for future primary science education that resulted from the large-scale project:

1. Good progress in science requires effective teaching,
2. Good primary science education is dependent upon good science leadership, and
3. Science is vital for everyone's future, so it must be accessible (Stubberfield & Barton, 2021, pp. 9–10).

These principles provide some guiding thoughts in terms of considering the future of primary science teacher education and the preparation of future primary teachers of science who are sufficiently resourced to effectively enact these goals.

This chapter takes a future-oriented approach to reflecting on the learnings uncovered over the book and highlighting the areas that still need to be considered in regard to primary science teacher education in Australia. In providing concluding commentary, insights will identify possible trends and patterns emerging in primary science teacher education, locally and internationally, with a focus on considering what future primary science teacher education is preparing graduates for and therefore what future-focused approaches and attributes may be required.

6.2   What Is the Future?

If we are to effectively prepare for the future, then some future forecasting is required. While this book has sought to think about science education as being much more than an employment pipeline, it is still important to give the future some consideration alongside an imagining of future workplace possibilities (OECD, 2018). Caution must be given, however, when framing education as a potential factor for success in obtaining and succeeding at work. As Buchanan and colleagues (2020) point out, while education is an important part of any policy landscape, it cannot compensate for the inadequacies of other policy areas that have caused and continue to cause, globally, widespread unemployment and under-employment. Instead, education can provide the foundational support required to enable an individual's cognitive, inter-personal and intrapersonal capacities to develop and flourish in ways that will equip them with the skills and knowledges necessary for the world of work. According to Buchanan et al. (2020), "while education is not 'the answer' to the challenges of the future of work, there can be no answer to the challenges without a quality education system" (p. 17).

A recent industry-university partnership between Deakin University (Australia), Griffith University (Australia) and Ford Motor Company resulted in the develop-ment of the *100 Jobs of the Future* research project (Tytler et al., 2019a) and an online tool known as the *Future Job Quiz* (Tytler et al., 2019b). This Australian-based collaboration aimed to examine the future of work in light of an increasingly technologically-driven society by deeply considering work trends and patterns within a changing societal demographic and environmental context, before engaging experts familiar with cutting edge developments in a range of fields to construct a picture of what the jobs in future workplaces might look like. The resultant 100 jobs draw on a knowledge base that is largely scientific, technological, and digital in nature but will require personal attributes such as interpersonal skills, creativity, and imagination to bring these cognitive elements to life (Tytler et al.,). Examples of future jobs that have a science-emphasis include: de-extinction geneticist, genetics coach, cricket farmer, and virtual surgeon.

While the *100 Jobs of the Future* may seem somewhat whimsical, there are key messages to extract from the report and online resources that have significant implica-tions for science education. The most notable is a move away from science education being about teaching students to think in disciplinary ways to having a more inter-disciplinary focus. Interdisciplinarity still requires students to be supported in devel-oping a strong disciplinary base but combines this science-focused knowledge with knowledge from other areas alongside technological skills, creativity and personal attributes (e.g., flexibility, adaptability) (Tytler et al., 2019a, 2019b). Bull (2019) notes that as we transition to a post-carbon world, there will be numerous complex issues to grapple with and that science education will need to be ready to assist students in learning to hold multiple perspectives and embrace complexity. Given the seriousness of some of the looming issues, it would be understandable for students to start to despair and lose their sense of hope. Instead, the *100 Jobs of the Future* provides science education with a platform from which to build student agency and acknowledge that there will be future avenues in which they are positioned to make a difference. This signals a real shift from teaching science as separate disciplines (e.g., biology, chemistry, physics) to a vehicle for exploring current and future social and environmental problems (e.g., water literacy, alternative energy sources).

6.3   Future Approaches

With an increased sense of the future, the opportunities for science teachers and teacher educators in preparing students and teachers, respectively, for their future have gained some clarity. What requires further exploration now, however, are the approaches that will refine and redefine science education itself into the future.

6.3.1   Technology

Digital technologies are playing an increasingly important role in our lives and will no doubt continue to do so into the future. As a response, Australian educational systems have ramped up the integration of both design and digital technologies in the curriculum (see: ACARA, 2022c) and some state governments have mandated particular digital technologies, such as coding, to be integrated into the existing syllabus and taught in the classroom (see: Baker, 2018). To support these changes, significant resourcing has been directed towards the upskilling of the current teacher workforce to increase proficiency in this area (see: ACARA, 2021), for example, the introduction of 'STEM Champions' across primary schools in the north-eastern Australian state of Queensland (see Chap. 4). As part of accreditation processes, ITE programs, both primary and secondary, are expected to have at least one course focused on digital and design technology education for pre-service teachers (AITSL, 2015). Alongside this requirement, many ITE providers refer to technologies being integrated across the courses in their programs. It is worth noting that while an integrated approach makes sense as a way of contextualising technology use, the implicit teaching of technologies does run the risk of being 'invisible' to pre-service teachers and becoming 'buried' within the program or ultimately not taught at all (Rawlins & Kehrwald, 2014).

In thinking about future technology needs and the role of classroom instruction, substantial time and energy are being dedicated to encouraging Australian teachers to engage with digital technologies, such as robotics with a particular focus on coding, as part of their teaching practice (Baker, 2018). The scale of subsequent resourcing has been significant with the unfortunate reality being that many schools across the country now have thousands of dollars in hardware, such as drones, 3D printers, and various robots (e.g., Beebots, Spheros), sitting in cupboards and gathering dust (Bolton, 2020). Without a large-scale commitment to the upskilling of all teachers, not just those who are comfortable or enthusiastic about digital technologies, the uptake of meaningful and contextually relevant technology education in the classroom might be challenging (Vivian & Falkner, 2018). This education system-wide focus on hardware misses the real value of technology education and the development of the skills that will actually be needed in the future, which are the cognitive elements that inform how technology is manipulated for positive outcomes, namely computational, design, and systems thinking (Ferrari, 2012).

As so called 'digital natives', current pre-service teachers are certainly savvy in their use of technology on a day-to-day basis (Thompson, 2013). They are comfortable and confident in accessing different types of technology and are unafraid to problem shoot any difficulties. Conversely, primary science teacher educators, in general, may feel less positive about their capacity and capabilities in these areas (Vivian & Falkner, 2018). Regardless, fluency and familiarity with technology do not equate directly to competency in using technology appropriately and effectively as a pedagogical tool (Thompson, 2013). Primary science teacher education has a role to play in providing the foundational elements of technology education needed to enrich and enhance science learning and teaching (Fakherji, 2019). This shift will largely be achieved through future teachers of science gaining a grounding in the thinking that underpins the implementation of technology (e.g., design thinking) rather than focusing solely on the more technical elements (e.g., how to use a piece of hardware).

6.3.2 Global Sustainable Development Goals

In 2015, all United Nations (UN) Member States (193 countries and two non-member observer states: the Holy See and the State of Palestine) agreed to adopt the 2030 Agenda for Sustainable Development (UN, 2015). The intention of this agenda was to create a shared vision and detailed plan of action to lead to improvements for "people, planet, prosperity, peace, and partnership" (UN, 2015, p. 1). At the core of the action plan are 17 Sustainable Development Goals (SDGs), which are essentially the blueprint to achieving this vision with key outcomes and indicators of success (see: DESA, 2015). This call to action brings all countries, developed and developing, together in a global partnership that builds on decades of work led by the UN to create a path towards global sustainable development.

The annual SDG Progress Report (UN, 2022) for 2022 was released in July and illustrates a grim picture for the achievement of the 2030 Agenda for Sustainable Development plan due to multiple, ongoing and intersecting crises (e.g., pandemic, conflict, climate change).

If the SDGs are to be achieved as intended, it is critical that future decision-makers and key societal stakeholders understand the importance of these goals and the role they play in their achievement both personally and through their influence of others (Kleespies & Dierkes, 2022). Teachers and teacher educators are well placed to actively contribute towards making a difference through the SDGs. The influence of ITE programming on education for sustainable development (ESD) is not new with this particular body of knowledge spanning over 15 years (e.g., Summers et al., 2005). A particularly important finding from this corpus of work is that interdisciplinary approaches are key to leading ESD innovation and collaboration within ITE programs and school settings alike (Chisingui & Costa, 2020). In particular, science education and by extension science teacher education are well positioned to leverage the inherent skills, knowledges, and practices connected with science in contributing to action around the SDGs in education (Eloff et al., 2022).

The Smithsonian Science Education Centre (SSCC) is leading the way in ESD space with their *Smithsonian Science for Global Goals* project (SSCC, 2022). This project has developed comprehensive curriculum for use by educators broadly to support students aged 8 to 17 years around the world to develop science knowledge and skills that will assist them to understand some of the world's most pressing issues (e.g., biodiversity, water, pandemics, mosquito-borne disease) and to ultimately become change agents in their own communities through improved and informed decision-making. This focus is positioned against the backdrop of increased agency and voice amongst young people to express their thoughts around issues of concern (e.g., School Strike 4 Climate actions). Two key elements of this work, which can be translated into classrooms and ITE programming, are: (i) the use of an interdisciplinary approach to learning as a way of making sense of real-world problems, and (ii) the contextualisation of global issues back into local settings to better reflect community need, interest and priorities. Primary science teacher education can leverage these conditions within courses to enable future teachers of science to grapple with coming to understand the SDGs in this way and be more adequately prepared to support their own students with this work into the future.

6.3.3   Scientific Literacy

The Organisation for Economic Co-operation and Development (2002), reporting for the *Programme for International Student Assessment* (PISA), defines scientific literacy as "the capacity to use scientific knowledge, to identify questions, and to draw evidence-based conclusions in order to understand and help make decisions about the natural world and the changes made to it through human activity" (p. 1). With this concept being increasingly viewed as the primary goal of school science, there is widespread agreement that the purpose of science education should be developing scientifically literate citizens who are well positioned to navigate an increasingly complex and digital world (Howell & Brossard, 2021; Reiss, 2007). If the aim of science education is to develop scientific literacy, then there is a need to understand what it is that characterises the behaviours of a scientifically literate person.

As part of the call for a greater focus on developing scientific literacy in Australian schools, the seminal report *The Status and Quality of Teaching and Learning of Science in Australian Schools* identifies a number of attributes of a scientifically literate person. This list emphasised that scientifically literate people are:

- interested in and understand the world about them;
- able to identify and investigate questions and draw evidence-based conclusions;
- able to engage in discussions of and about science matters;
- sceptical and questioning of claims made by others; and
- able to make informed decisions about the environment and their own health and wellbeing (Goodrum et al., 2001, p. 7).

Despite the report being two decades old, these attributes continue to have currency and emphasise scientifically literate citizens as being curious, questioning and having the capacity to engage with science in ways that allow them to view the world scientifically (Ashbrook, 2020). Rather than being solely discipline-based, this perspective of scientific literacy focuses on the development of a more generic set of skills that would be of assistance in dealing with scientific issues, ideas, and phenomena that impact on daily life. In broad agreement, Norris and Phillips' (2003) examination of the literature resulted in a list of factors identifying scientific literacy, which included: the desire to be an independent, lifelong learner of science; to have a willingness to engage with science ideas; and the ability to interpret and construct science texts. Their review argues that in fostering scientific literacy, equal attention should be given to both students' orientations to science and their abilities regarding their understanding and application of scientific ideas.

While these different interpretations contribute to understandings of scientific literacy, greater clarity can be brought to this construct through primary science teacher education. In bringing meaning to the literature concerning scientific literacy, Roberts (2007) referred to two visions representing the continuum of understanding. Vision 1, the traditional end of the scale, focuses on the processes and products of science itself, therefore examining literacy from within the practice of science. Whereas, Vision 2, adopting a more socio-scientific approach, examines the scientific components of situations that students are likely to be faced with in their daily lives, suggesting literacies that connect with science-related situations. The learning of science can be viewed as an active and adaptive process rather than simply leading to resolved conceptual end points. The literacies of science should therefore be considered as an important teaching focus (Tytler, 2007). Science teaching should promote the development of scientific literacy and assist students in the process of actively making informed decisions about science-based issues impacting on them at a public and on personal levels (Laugksch, 2000). While scientific literacy has gained significant traction in science education research, it is yet to be articulated and implemented regularly in classroom science learning and teaching practices. Primary science teacher education has a role to play in shifting this deficit perspective.

6.4 Future Attributes

The role of primary science teacher education is to equip future teachers of science with the knowledges and skills of science that move beyond being discipline-specific. Primary science teacher education creates the conditions to support pre-service teachers to understand science as an evidence-based way of making sense of the world and that from this platform other knowledges can be drawn upon to solve complex, real-world problems. In achieving this personal growth, this section explores some considerations for the future as primary pre-service teachers seek to implement what they have learnt from their primary science teacher education in their classroom practice.

6.4.1 Notions of Identity Formation

Fitzgerald's (2012) doctoral research drew out practices that are characteristic of effective teachers of primary science. In summary, there were three elements underpinning what it means to be a teacher of primary science: (i) having an understanding of cohort needs and contextual factors; (ii) engagement with hands-on, inquiry-based approaches to science learning and teaching; and (iii) opportunities for students to talk about and represent science in a range of ways (Fitzgerald, 2012). What this work and the research of others researching in primary science education (e.g., Kane & Varelas, 2016) discovered that for these 'best' practices to be present, primary school teachers are required to construct identities that enable them to identify as becoming or being a 'teacher of science'. This identity construction is intimately intertwined with the learning process, which in this context includes learning about science ideas, pedagogical challenges, instructional approaches, and curricular decision making (Kane & Varelas, 2016).

At its most basic level, supporting primary teachers to develop their own teacherly identities to include being a teacher of science matters because it means that their students will be engaged in science learning and have a significant opportunity to develop their own identities as learners of science (Kane & Varelas, 2016). If a primary teacher has confidence and competence that science learning and teaching is part of their teaching repertoire, then they are more likely to enact meaningful science education opportunities in their classrooms (Fitzgerald & Smith, 2016). If the role of science education is indeed to support the development of a scientifically literate citizenry, then making a positive start in primary school is of critical importance (Nilsson, 2011). The pathway to positive experiences of science starts in the early years of schooling, but this is only possible if primary school teachers are appropriately equipped through their own experiences of primary science teacher education to make this possible (Ginns & Watters, 1999).

The context in which teachers work is complex and dynamic. Therefore, equipping graduate teachers with the knowledge and skills needed to engage with the intellectual and practical work associated with ambitious and equitable science teaching practices as well as the drive to become a life-long learner and leader in science education is incredibly important (Windschitl & Calabrese Barton, 2016). Unfortunately, the past science and school experiences of future primary teachers can elicit significant anxiety and create a barrier to becoming a teacher of science. This tension has been further reinforced in the Australian context with most ITE providers removing the completion of at least one senior secondary science (e.g., biology, chemistry, physics) as a prerequisite to admission to ITE programs. In using identity construction as a framework, primary science teacher education has an opportunity to positively shift this narrative (Zembal-Saul, 2016). In circling back to Fitzgerald's (2012) research, primary science teacher education courses need to be created with cohort needs and contextual features in mind before identifying the most appropriate ways to engage pre-service teachers positively in science as learners. Alongside this learning, primary science teacher education has a key supporting role in eliciting conversations around how ITE experiences can be translated into future science teaching practices and therefore further enhance the development of teacher identity (Avraamidou, 2014).

6.4.2 Considerations Around Agency

One of the challenges faced in primary science teacher education is diminished opportunities for pre-service teachers to observe and experience the learning and teaching of science in primary school classrooms (Fitzgerald et al., 2020). While they may graduate from their ITE program with increased confidence and competence to embrace their identity as a teacher of science (Kazempour, 2013), pre-service teachers are often doing so with little to no experience of applying their knowledge, skills, and practices directly with primary-aged students, though there are some noteworthy exceptions (see: Fitzgerald, 2020; Hobbs et al., 2015) (see Chap. 3 for further exploration and explanation). This reality, unfortunately, tends to spill over into classroom practice with graduates experiencing a lack of agency in implementing science education due to a variety of factors, including: a lack of support from other teachers and/or school leadership; a misalignment with school and/or departmental strategic goals; and a grappling with how to make this happen within a 'crowded curriculum' (Severance & Krajcik, 2018).

At its simplest level, agency is understood as a capacity to act. This explanation can be extrapolated to the work of teachers by defining teacher agency, more specifically, as the capacity to act "purposefully and constructively to direct their professional growth and contribute to the growth of education quality" (Guoyuan, 2020, p. 2). Having a sense of agency in educational work matters as teachers play a central role in the development and implementation of school-based curriculum, which intimately links with notions of quality, accountability and the enactment of policy reform (Priestley et al., 2015). On a deeper and more personal level, to lack agency as a teacher can be disheartening and dispiriting, which can have long-term impacts on both teacher identity and classroom practices (Priestley et al., 2015). On the flipside, having a strongly defined sense of belonging and purpose can be motivating and energising for the teacher as well as students and colleagues alike (Guoyuan, 2020).

Primary science teacher education equips and empowers graduates to be future teachers of science. As part of this work, there is a responsibility to provide pre-service teachers with the resources and capacity to be able to maintain this sense of agency around science learning and teaching when the supports of an ITE program are no longer present. Two particular approaches that can be useful in enhancing teacher agency are: professional learning opportunities and professional learning communities (Earle & Bianchi, 2022). A key way to achieve both of these outcomes is through membership of teacher professional associations, which have an important role to play in supporting teachers in maintaining their currency and their connections often within a particular learning area (e.g., science) or educational focus (e.g., inclusive education) (NSHSS, 2019). In the Australian context, there are national and state-based science teacher associations, which all teachers can become a member of for a small fee (see: ASTA, 2022). Primary science teacher education courses themselves are a rich networking opportunity and pre-service teachers should be encouraged to stay connected with this particular group of future colleagues. As mentioned in Chap. 4, science specialist teachers who had collaborations in the form of professional learning networks were highly successful in their work compared with those who didn't and as a result felt professionally isolated.

6.4.3   Professional Learning

Learning about science and science education doesn't end with the completion of a science education course (Bull, 2016). In fact, the finishing of an ITE degree should signal the start of being dedicated to lifelong learners (Yang et al., 2015). Access to meaningful and relevant learning opportunities to upskill and improve is critical for future teachers to be successful in this pursuit of knowledge (Li, 2022). Commonly, primary teachers of science tend to strike out on their own to increase their capacity and broaden their science education skillset (Bull, 2016). This individual pursuit of improvement is often a reflection of their own lack of confidence and competence about science and science education (Loughran, 2014). And while, in the Australian context, there is generally access to a range of activities to support primary science learning and teaching, these opportunities are often disconnected from each other, the pressures of the school system, and often classroom realities (Fitzgerald & Smith, 2016).

At the heart of why quality professional learning in science education matters is its criticality to supporting and enhancing ongoing effective science instruction in schools, particularly for primary teachers of science (National Research Council, 2007). In the professional learning stakes, it is important to recognise that teachers of primary science learn best when exposed to relevant and meaningful hands-on opportunities to learn science, just like their own students (Professional Learning & Development Advisory Group, 2014). Alongside lived science experiences, primary teachers of science learn and develop best as effective teachers when they not only understand the content they are presenting to their students, but also grasp the pedagogy behind how their students learn science (Mupa & Chinooneka, 2015). In understanding these elements, it is important to recognise that science education professional learning must guide primary teachers of science "in delivering relevant, useful, and meaningful science into very crowded, complex, and contested classrooms" (Bell & Sexton, 2018, p. 119).

The following quote reflects on a perceived disjuncture between science teacher education (in general rather than with a specific primary focus) and the realities of what skills and knowledges future teachers of science need to enact effective practices in contemporary classrooms. "These [professional learning] needs represent a significant change from what virtually all active teachers learned in college and what most colleges teach aspiring teachers today" (National Research Council, 2007, p. 7). The existence of an evidence base that suggests many primary teachers are not prepared to enact quality science instructions (e.g., reduced instructional time, reduced confidence) in their classrooms is a tough blow for primary science teacher educators. Rather than being disheartened, this points to a significant opportunity for primary science teacher education courses to be grappling with what the needs are in the current schooling climate and to ensure future teachers are prepared to meet these needs. Primary science teacher educators, however, also have a responsibility to be aspirational and to create the conditions that encourage pre-service teachers to be future-focused as they prepare to become teachers of science (Wickham, 2015).

6.5 Future Teachers

In looking to the future in relation to the role of education, this section considers the role of future teachers of science in this imagined reality and what contribution primary science teacher education can make in preparation.

6.5.1 The Role of Future Teachers of Science

If there are some nuggets of truth in the above predictions about the future of work, then some introspection is needed to examine how students are being educated to navigate these impending challenges. Globally, concerns are being raised about the readiness and ability of school systems to address these changing needs in relation to the purpose of education (Krishnan, 2020). Significant equity and access imbalances exist that translate to students not having access to the information they need to prepare them for success in the job market or to make sense of the complex challenges they will face in their lives (Krishnan, 2020). Concerningly, UNICEF (2017) estimates that a staggering 72 million primary school-aged children are not in school and that in a considerable knock-on 750 million adults are illiterate and do not have the ability to improve their own, let alone their children's living conditions. Conversely, students who do have access to education are often constrained by curriculum that emphasises didactic instruction and the 'siloing' of learning areas (Mupa & Chinooneka, 2015).

A prominent science journal connected with educators across the world and asked them to respond to the following question: What one improvement could your country make to its current education system to prepare students to face future challenges? (Pietrzak, 2018). The subsequent responses fell into two key categories: connection with nature and interdisciplinary collaboration (Pietrzak, 2018). In terms of school-based practice, these insights suggest a foregrounding of approaches to learning and teaching that are place-based and value opportunities to draw on the skills and knowledges from across disciplines. While this translation doesn't seem revelationary or revolutionary, the structural constraints imposed by educational systems continue to be problematic by acting to constrain rather than catalyse change. In the Australian context, the curriculum (see: ACARA, 2022a) in its current form doesn't provide teachers with scope or inspiration to easily contextualise or integrate the eight discipline-based learning areas. To be fair to the curriculum developers, attempts at interdisciplinarity are present in the inclusion of three cross-curriculum priorities, (i) Aboriginal and Torres Strait Islander Histories and Cultures, (ii) Asia and Australia's Engagement with Asia, and (iii) Sustainability (ACARA, 2022b), which in reality can be lost in practice through learning and teaching experiences that can be perceived as disconnected or irrelevant to students. This often extends to, or is even worse in, senior school where research has documented priorities such as sustainability "fall by the wayside" (e.g., Tomas et al., 2020, 2022).

Teachers of science are in the box seat to make a difference through education to the development and preparation of future-ready citizens. Science education lends itself to practices that are contextualised in the local environment and draw on interdisciplinary skills and knowledge to solve complex problems (Buxton & Provenzo, 2012). On paper, instigating such a change through the use of so-called 'innovative pedagogies' (e.g. integration of virtual reality, use of 3D printing, design-based thinking) seems very possible, however, the challenge lies in the preparedness of teachers of science to make this happen (see Chap. 3 for more detail around pedagogies and practices in primary science teacher education). A balancing act is required to work within the expectations of the curriculum and structures of a school, while empowering students to use science as a lens for making sense of their world and to be willing to be guided by the 'teachable moments' that will subsequently and unpredictably unfold. For many teachers, this will require an act of bravery, of moving outside of comfort zones into the discomfort of the learning zone.

6.5.2 The Contribution of Primary Science Teacher Educators

Primary science teacher educators are critical in equipping future teachers of science with the skills and knowledge to effectively survive, and potentially thrive, in places of discomfort (Fitzgerald & Smith, 2016). Essentially, primary science teacher education courses have an important role to play in planting the 'seeds' of possibility by moving pre-service teachers from their past experiences and perceptions of what science education is to re-imagining what it could be (Tytler, 2007) (see Chap. 2 for more detail around the purpose of primary science teacher education). This task is a challenging one as not only is there the social and cultural 'baggage' attached to science education, but there is often a lack of visibility and encouragement in schools from leaders and teachers alike to engage differently with science learning and teaching (Jones & Burrell, 2022). Subsequently, what emerges in the minds of pre-service teachers is a tension between 'theory' and how ITE providers view education and 'practice' and the ways in which teachers engage with education. The widening of this divide potentially drives a wedge between evidence-based innovation and traditional approaches to science learning and teaching.

Circling back to Wellcome Trust's work around the three principles for future primary science education (Stubberfield & Barton, 2021) is a potentially useful mechanism for further unpacking the role of primary science teacher educators in preparing future teachers. Primary science education courses have a responsibility to challenge pre-service teachers in their understanding of effective teaching (Fitzgerald, 2012). In initiating this challenge, primary science teacher educators need to draw on a combination of their understanding of current classroom practices, knowledge of cutting-edge research, and strategies for bringing these two elements together in meaningful ways for pre-service teachers through lived experiences (Eley, 2022). Of particular importance and relevance to primary science education is the creation of opportunities for all students to find a way to connect and access science (Fitzgerald, 2012). Primary science teacher educators typically seek to achieve this through providing the conditions that enable pre-service teachers to see their world through the 'wonder' of science to provoke a sense that science is everywhere and can be a useful way in which to make sense of the world (Gilbert & Byers, 2017). At a pre-service level, the promotion of good science leadership is slightly more challenging (see Chap. 4 for more detail around preparing primary teachers as science leaders), but the development of more confident and competent future teachers of primary science are certainly steps in a positive direction (Fitzgerald, 2020).

Primary science teachers educators can't predict the future, but they can use trends and patterns to make educated guesses about the science learning and teaching needs of future teachers. There is certainly a shift away from primary science teacher education being about equipping pre-service teachers with only content knowledge to much more nuanced considerations of the types of pedagogical content knowledge they will require (Timostsuk, 2015). The shift away from a content focus is also in recognition that it is not entirely possible to teach science conceptual understandings to pre-service teachers in the limited timeframes of semester-long courses, but that instead they should be equipped with the skills and knowledge in how to upskill their discipline knowledge as needed and know how to access reputable sources to assist (Tekin et al., 2020). Equally, primary science teacher educators have a responsibility for instilling in future teachers of science a sense of wonder about the world to assist them in staying curious and open to learning alongside their students (Gilbert & Byers, 2017).

6.6   Conclusion: Looking to the Future Classrooms and Beyond

Given the global challenges of the last few years, the importance of quality science education as a way of making sense of the world and solving complex problems has been clearly cemented in the psyche of the general public and policy makers alike. To revisit the quote from Stubberfield and Barton (2021), "high quality primary science teaching ensures that young people will be better prepared for their futures" (p. 4). Primary teachers of science are positioned front and centre to make a difference to not only the science/STEM-focused workforce pipeline but, and perhaps more importantly, to equip future citizens with the skills and knowledges to navigate the uncertain world that lies ahead. It is, therefore, an exciting time for primary science teacher education with a readiness for innovation and change in the wind, in relation to science education specifically but also the higher education sector more broadly. This excitement is slightly tempered by the somewhat daunting nature of the role. To be a primary science teacher educator preparing future-focused teachers of science is a balancing act: to excite pre-service teachers for the possibilities inherent in science education and equip them for the realities of the classroom and school system while drawing on a mix of innovation and theory alongside pragmatics and practice.

It is possible that primary science teacher education has a harder lift than some of the other offerings within ITE programming due to some of the perceptions and past experiences that pre-service teachers bring with them. It is also challenging because some of these perceptions and past experiences have a tendency to be re-perpetuated in some classroom and school contexts by teacher and school leaders, who themselves are lacking in their own confidence and competence around science education. Primary science teacher education has a direct role in breaking this cycle by working with pre-service teachers to develop their identities as future teachers of science alongside a sense of agency and strong professional networks to propel them along as a leader in science learning and teaching into their graduate years and beyond. While primary science teacher educators are more than up to this task, they also face their own challenges in terms of their own currency of knowledge and practice in the ever-evolving fields of science and science education. It is also increasingly important that teacher educators look to their own motivations and values to understand what drives them as a science teacher educator and what they truly believe the purpose of primary science education to be.

So, where is primary science teacher education headed? This chapter, and by and large this book, proposes an optimistic way forward. The conditions are certainly ripe for change and the importance of preparing future teachers of science, particularly for primary school contexts, is indisputable. In guiding these next steps for primary science teacher educators, three considerations are posed:

1. Quality science education is required to prepare students for an uncertain future;
2. Classroom-ready and future-focused teachers of science are needed to equip students with the skills and knowledges required to be science literate citizens; and

3. Primary science teacher education has a critical role to play in providing future teachers of science not only with the theory and practice to inform their roles, but the confidence, agency, and conviction to make sure that meaningful and relevant science learning and teaching are front and centre in their classrooms.

References

Ashbrook, P. (2020). Becoming scientifically literate. *Science and Children, 57*(8).

Australian Curriculum, Assessment and Reporting Authority [ACARA]. (2021). *Digital technologies in focus.* https://www.australiancurriculum.edu.au/resources/digital-technologies-in-focus/

Australian Curriculum, Assessment and Reporting Authority [ACARA]. (2022a). *F-10 curriculum overview.* https://v9.australiancurriculum.edu.au/f-10-curriculum/f-10-curriculum-overview/

Australian Curriculum, Assessment and Reporting Authority [ACARA]. (2022b). *Cross-curriculum priorities.* https://v9.australiancurriculum.edu.au/f-10-curriculum/f-10-curriculum-overview/cross-curriculum-priorities

Australian Curriculum, Assessment and Reporting Authority [ACARA]. (2022c). *Design and technologies (V8.4).* https://www.australiancurriculum.edu.au/f-10-curriculum/technologies/design-and-technologies/

Australian Institute of Teaching and School Leadership [AITSL]. (2015). *Accreditation of initial teacher education programs in Australia.* https://www.aitsl.edu.au/docs/default-source/national-policy-framework/accreditation-of-initial-teacher-education-programs-in-australia.pdf?sfvrsn=e87cff3c_48

Australian Science Teachers Association [ASTA]. (2022). *Promoting our profession, enriching science teaching.* https://www.asta.edu.au

Avraamidou, L. (2014). Studying science teacher identity: Current insights and future research directions. *Studies in Science Education, 50*(2), 145–179. https://doi.org/10.1080/03057267.2014.937171

Baker, J. (2018, August 21). Coding to be mandatory in primary, early high school. *Sydney Morning Herald.* https://www.smh.com.au/national/nsw/coding-to-be-mandatory-in-primary-early-high-school-20180817-p4zy5d.html

Bell, S. E., & Sexton, S. S. (2018). Science education professional development for primary/elementary teachers: A tale of two systems. *Science Education International, 29*(2), 117–123. https://doi.org/10.33828/sei.v29.i2.7

Bolton, R. (2020, April 20). Australian schools tech ready, but not teachers. *Australian Financial Review.* https://www.afr.com/work-and-careers/education/australian-schools-tech-ready-but-not-teachers-20200416-p54kd4

Buchanan, J., Allais, S., Anderson, M., Calvo, R. A., Peter, S., & Pietsch, T. (2020). *The futures of work: What education can and can't do.* UNESCO: Education Sector. https://unesdoc.unesco.org/ark:/48223/pf0000374435

Bull, A. (2016). *Developing primary science teacher expertise: Thinking about the system.* New Zealand Council for Educational Research. https://www.nzcer.org.nz/system/files/Primary%20Science%20Report.pdf

Bull, A. (2019). *Thinking about science education for the future.* Science Learning Hub. https://www.sciencelearn.org.nz/resources/2890-thinking-about-science-education-for-the-future

Buxton, C., & Provenzo Jr., E. F. (2012). https://sk.sagepub.com/books/place-based-science-teaching-and-learning

Chisingui, A. V., & Costa, N. (2020). Teacher education and sustainable development goals: A case study with future biology teachers in an Angolan Higher Education Institution. *Sustainability, 12*(3344), 1–14. https://doi.org/10.3390/su12083344

DESA. (2015). *The 17 goals*. United Nations Sustainable Development. https://sdgs.un.org/goals#goals

Duraiappah, A. K., van Atteveldt, N. M., Borst, G., Bugden, S., Ergas, O., Gilead, T., Gupta, L., Mercier, J., Pugh, K., Singh, N. C., & Vickers, E. A. (Eds.). (2022). *Reimagining education: The international science and evidence-based education assessment*. UNESCO. https://doi.org/10.56383/IQAX5928

Earle, S., & Bianchi, L. (2022). What role can professional learning frameworks play in developing teacher agency in subject leadership in primary science? *Professional Development in Education, 48*(3), 462–475. https://doi.org/10.1080/19415257.2021.1942142

Eley, A. (2022). Different approaches to teaching primary science the role of the teacher is what makes the difference. *Primary Science, 171,* 7–10.

Eloff, I., Mathabathe, K., Agostini, E., & Dittrich, A.-K. (2022). Teaching the global goals: Exploring the experiences of teacher education in an online environment through vignette research. *Environmental Science Proceedings, 15*(5), 1–9. https://doi.org/10.3390/environscipro oc2022015005

Fakherji, W. Z. (2019). Teachers' use of technology in science supports student knowledge. *Journal of Research in Curriculum, Instruction and Education Technology, 5*(1), 135–158. https://doi. org/10.21608/jrciet.2019.31979

Ferrari, A. (2012). *Digital competence in practice: An analysis of frameworks*. Joint Research Centre. https://actic.gencat.cat/web/.content/01_informacio/documents/arxius/dc_in_ practice._analysis_of_frameworks.pdf

Fitzgerald, A. (2012). *Science in primary schools: Examining the practices of effective primary science teachers*. Sense Publishers.

Fitzgerald, A. (2020). Out in the field: The impact of school-based experiences on primary pre-service teachers' confidence and competence to teach science. *International Journal of Science Education, 290–309.* https://doi.org/10.1080/09500693.2019.1710618

Fitzgerald, A., Pressick-Kilborn, K., & Mills, R. (2020). Primary teacher educators' practices in and perspectives on inquiry-based science education: Insights into the Australian landscape. *Education 3–13: International Journal of Primary, Elementary and Early Years Education, 49*(3), 344–356. https://doi.org/10.1080/03004279.2020.1854962

Fitzgerald, A., & Smith, K. (2016). Science that matters: Exploring science learning and teaching in primary schools. *Australian Journal of Teacher Education, 41*(4), 64–78. https://doi.org/10. 14221/ajte.2016v41n4.4

Gilbert, A., & Byers, C. C. (2017). Wonder as a tool to engage preservice elementary teachers in science learning and teaching. *Science Education, 101*(6), 907–928. https://doi.org/10.1002/sce. 21300

Ginns, I. S., & Watters, J. J. (1999). Beginning elementary school teachers and the effective teaching of science. *Journal of Science Teacher Education, 10*(4), 287–313. https://doi.org/10.1023/A: 1009442125203

Goodrum, D., Hackling, M., & Rennie, L. (2001). *The status and quality of teaching and learning of science in Australian schools*. Department of Education, Training and Youth Affairs. http:// www.dest.gov.au/schools_publications

Guoyuan, S. (2020). Teacher agency. In M. A. Peters (Ed.), *Encyclopedia of teacher education* (pp. 1–5). Springer. https://doi.org/10.1007/978-981-13-1179-6_271-1

Hobbs, L., Chittleborough, G., Jones, M., Kenny, J., Campbell, C., Gilbert, A., Redman, C., & King, J. (2015). *School-based pedagogies and partnerships in primary science teacher education: The Science Teacher Education Partnerships with Schools (STEPS) project* [Report]. Office for Learning and Teaching. https://ltr.edu.au/resources/ID12-2412_Hobbs_Final%20report_2015. pdf

Howell, E. L., & Brossard, D. (2021). (Mis)informed about what? What it means to be a science-literate citizen in a digital world. *PNAS, 118*(15). https://doi.org/10.1073/pnas.1912436117

Jones, T. R., & Burrell, S. (2022). Present in class yet absent in science: The individual and societal impact of inequitable science instruction and challenge to improve science instruction. *Science Education, 106*(5), 1032–1053. https://doi.org/10.1002/sce.21728

Kane, J. M., & Varelas, M. (2016). Elementary school teachers constructing teacher-of-science identities: Two communities of practice coming together. In L. Avraamidou (Ed.), *Studying science teacher identity: Theoretical, methodological and empirical explorations* (pp. 177–195). Springer. https://doi.org/10.1007/978-94-6300-528-9_9

Kazempour, M. (2013). I can't teach science! A case study of an elementary pre-service teachers' intersection of science experiences, beliefs, attitude, and self-efficacy. *International Journal of Environmental & Science Education, 9*, 77–96.

Khamis, T., Naseem, A., Khamis, A., & Petrucka, P. (2021). The COVID-19 pandemic: A catalyst for creativity and collaboration for online learning and work-based higher education systems and processes. *Journal of Work-Applied Management, 13*(2). https://doi.org/10.1108/JWAM-01-2021-0010

Kleespies, M. W., & Dierkes, P. W. (2022). The importance of the sustainable development goals to students of environmental and sustainability studies: A global survey in 41 countries. *Humanities and Social Sciences Communications, 9*(218), 1–9. https://doi.org/10.1057/s41599-022-01242-0

Krishnan, K. (2020). *Our education system is losing relevance. Here's how to unleash its potential.* World Economic Forum. https://www.weforum.org/agenda/2020/04/our-education-system-is-losing-relevance-heres-how-to-update-it/

Laugksch, R. C. (2000). Scientific literacy: A conceptual overview. *Science Education, 84*(1), 71–94. https://doi.org/10.1002/(SICI)1098-237X(200001)84:1%3c71::AID-SCE6%3e3.0.CO;2-C

Leonardi, S., Tyers, C., Lamb, H., Milner, C., Howe, P., Hansel, M., & Spong, S. (2021). *The impact of COVID-19 on primary science education.* Wellcome Trust. https://cms.wellcome.org/sites/default/files/2021-09/the-impact-of-covid-19-on-primary-science-education.pdf

Li, L. (2022). Reskilling and upskilling the future-ready workforce for Industry 4.0 and beyond. *Information Systems Frontiers*, 1–16. https://doi.org/10.1007/s10796-022-10308-y

Loughran, J. J. (2014). Developing understandings of practice: Science teacher learning. In N. G. Lederman & S. K. Abell (Eds.), *Handbook of research on science education* (Vol. 2, pp. 811–829). Routledge.

Marmolejo, F. J., & Groccia, J. E. (2022). Reimagining and redesigning teaching and learning in the post-pandemic world. *New Directions for Teaching and Learning, 169*, 21–37. https://doi.org/10.1002/tl.20480

Mupa, P., & Chinooneka, T. I. (2015). Factors contributing to ineffective teaching and learning in primary schools: Why are schools in decadence? *Journal of Education and Practice, 6*(19), 125–133.

National Research Council. (2007). *Taking science to school: Learning and teaching science in Grades K-8.* The National Academies Press.

National Society of High School Scholars [NSHSS]. (2019). *5 reasons why it's important for educators to join professional organizations.* https://www.nshss.org/blog/5-reasons-why-it-s-important-for-educators-to-join-professional-organizations-1/

Nilsson, P. (2011). Why does scientific literacy matter in primary schools? Reflections on the OLGC experience. In J. J. Loughran, K. Smith, & A. Berry (Eds.), *Scientific literacy under the microscope* (pp. 127–137). Brill. https://doi.org/10.1007/978-94-6091-528-4_14

Norris, S. P., & Phillips, L. M. (2003). How literacy in its fundamental sense is central to scientific literacy. *Science Education, 87*(2), 224–240. https://doi.org/10.1002/sce.10066

Organisation for Economic Cooperation and Development. (2002). *Education at a glance.* Organisation for Economic Cooperation and Development. https://www.oecd.org/education/skills-beyond-school/educationataglance2002-home.htm

Organisation for Economic Cooperation and Development. (2018). *The future of education and skills: Education 2030*. Organisation for Economic Cooperation and Development. https://www.oecd.org/education/2030/E2030%20Position%20Paper%20(05.04.2018).pdf

Pietrzak, B. (2018). Education for the future. *Science, 360*(6396), 1409–1412. https://doi.org/10.1126/science.aau3877

Priestley, M., Biesta, G., & Robinson, S. (2015). Teacher agency: What is it and why does it matter? In R. Kneyber & J. Evers (Eds.), *Flip the system: Changing education from the bottom up*. Routledge. https://doi.org/10.4324/9781315678573-15

Professional Learning and Development Advisory Group. (2014). *Report of the professional learning and development advisory group*. http://www.services.education.govt.nz/pld/backgr ound/%20further-reading

PwC. (2020). *Where next for tertiary education? How the COVID-19 crisis can be the catalyst to reboot towards a stronger sector*. PricewaterhouseCoopers Consulting. https://www.pwc.com.au/government/where-next-for-tertiary-education.pdf

Rawlins, P., & Kehrwald, B. (2014). Integrating educational technologies into teacher education: A case study. *Innovations in Education and Teaching International, 51*(2), 207–217. https://doi.org/10.1080/14703297.2013.770266

Reiss, M. J. (2007). What should be the aim(s) of school science education? In D. Corrigan, J. Dillion, & R. Gunstone (Eds.), *The re-emergence of values in science education* (pp. 13–28). Springer. https://doi.org/10.1163/9789087901677_004

Roberts, D. A. (2007). Scientific literacy/science literacy. In S. K. Abell & N. G. Lederman (Eds.), *Handbook of research on science education* (pp. 729–780). Lawrence Erlbaum Associates.

Severance, S., & Krajcik, J. (2018, June 23–27). Examining primary teacher expertise and agency in the collaborative design of project-based learning innovations. In *Proceedings of the 13th international conference of the learning sciences (ICLS)*.

SSCC. (2022). *Smithsonian science for global goals*. https://ssec.si.edu/global-goals

Stubberfield, L., & Barton, T. (2021). *Primary science education beyond 2021: What next?* Wellcome Trust. https://cms.wellcome.org/sites/default/files/2021-11/Primary-science-educat ion-beyond-2021.pdf

Summers, M., Childs, A., & Corney, G. (2005). Education for sustainable development in initial teacher training: Issues for interdisciplinary collaboration. *Environmental Education Research, 11*(5), 623–647. https://doi.org/10.1080/13504620500169841

Tasquier, G. (2022). Scientific literacies for change making: Equipping the young to tackle current societal challenges. *Frontiers in Education, 7*, 1–20. https://doi.org/10.3389/feduc.2022.689329

Tekin, N., Aslan, O., & Yilmaz, S. (2020). Improving pre-service teachers' content knowledge and argumentation quality through socio-scientific issues-based modules: An action research study. *Journal of Science Learning, 4*(1), 80–90. https://doi.org/10.17509/jsl.v4i1.23378

TEQSA. (2021). *Forward impact of COVID-19 on Australian higher education* [Press release]. Australian Government. https://www.teqsa.gov.au/sites/default/files/forward-impact-of-covid-19-on-australian-higher-education-report.pdf?v=1635904356

Thompson, P. (2013). The digital natives as learners: Technology use patterns and approaches to learning. *Computers & Education, 65*, 12–33. https://doi.org/10.1016/j.compedu.2012.12.022

Timostsuk, I. (2015). Domains of science pedagogical content knowledge in primary student teachers' practice experiences. *Procedia—Social and Behavioural Sciences, 197*, 1665–1671. https://doi.org/10.1016/j.sbspro.2015.07.217

Tomas, L., Mills, R., & Gibson, F. (2022). 'It's kind of like a cut and paste of the syllabus': A teacher's experience of enacting the Queensland earth and environmental science syllabus, and implications for education for sustainable development. *Australian Educational Researcher (AER), 49*(2), 445–461. https://doi.org/10.1007/s13384-021-00439-7

Tomas, L., Mills, R., Rigano, D., & Sandhu, M. (2020). Education for sustainable development in the senior earth and environmental science syllabus in Queensland, Australia. *Australian Journal of Environmental Education, 36*(1), 44–62. https://doi.org/10.1017/aee.2020.7

Tytler, R. (2007). *Re-imagining science education: Engaging students in science for Australia's future*. Australian Council for Educational Research.

Tytler, R., Bridgstock, R., White, P., Mather, D., McCandless, T., & Grant-Iramu, M. (2019a). *100 jobs of the future*. Deakin University. https://100jobsofthefuture.com/report/

Tytler, R., Bridgstock, R., White, P., Mather, D., McCandless, T., & Grant-Iramu, M. (2019b). *What will the future work look like for you?* [Quiz]. https://100jobsofthefuture.com/quiz/

UN. (2015). *Transforming our world: The 2030 agenda for sustainable development* (A/RES/70/1). https://sdgs.un.org/sites/default/files/publications/21252030%20Agenda%20for%20Sustainable%20Development%20web.pdf

UN. (2022). *The sustainable development goals report: 2022*. https://unstats.un.org/sdgs/report/2022/The-Sustainable-Development-Goals-Report-2022.pdf

UNESCO. (2022). *COVID-19 education response: Dashboards on the global monitoring of school closures caused by the COVID-19 pandemic*. Institute for Statistics. https://covid19.uis.unesco.org/global-monitoring-school-closures-covid19/

UNICEF. (2017). *Literacy rates continue to rise from one generation to the next: Fact sheet no.45* (FS/2107/LIT/45). UNESCO Institute for Statistics. http://uis.unesco.org/sites/default/files/documents/fs45-literacy-rates-continue-rise-generation-to-next-en-2017_0.pdf

Vivian, R., & Falkner, K. (2018, October 4–6). A survey of Australian teachers' self-efficacy and assessment approaches for the K-12 digital technologies curriculum [Paper presentation]. In *Proceedings of the 13th workshop in primary and secondary computing education*. https://doi.org/10.1145/3265757.3265762

Wickham, C. B. (2015). *A call for mindful teaching: Cultivating pre-service teachers' dispositions* [Doctoral dissertation, College of Saint Mary]. https://www.csm.edu/sites/default/files/Wickham_Dissertation.pdf

Windschitl, M., & Calabrese Barton, A. (2016). Rigor and equity by design: Locating a set of core teaching practices for the science education community. In D. H. Gitomer & C. A. Bell (Ed.), *Handbook of Research on Teaching* (5th ed., pp. 1099–1158). American Educational Research Association. https://doi.org/10.3102/978-0-935302-48-6_18

Yang, J., Schnelle, C., & Roche, S. (2015). *The role of higher education in promoting lifelong learning*. UNESCO Institute for Lifelong Learning. https://files.eric.ed.gov/fulltext/ED564050.pdf

Zembal-Saul, C. (2016). Implications of framing teacher development as identify construction for science teacher education research and practice. In L. Avraamidou (Ed.), *Studying science teacher identity: Theoretical, methodological and empirical explorations* (pp. 321–331). Springer. https://doi.org/10.1007/978-94-6300-379-7_15